The Green Mountain Riflemen

The Green Mountain Riflemen

Company F, First United States
Sharpshooters in the American Civil War

William Y. W. Ripley

LEONAUR

The Green Mountain Riflemen: Company F, First
United States Sharpshooters in the American Civil War
by William Y. W. Ripley

First published under the title
Vermont Riflemen in the War For the Union, 1861 To 1865
A History of Company F, First United States Sharp Shooters

Published by Leonaur Ltd

Material original to this edition and this editorial selection
copyright © 2011 Leonaur Ltd

ISBN: 978-0-85706-535-3 (hardcover)
ISBN: 978-0-85706-536-0 (softcover)

http://www.leonaur.com

Publisher's Notes

Contents

Abate the edge of traitors, gracious Lord.
That would reduce these bloody days again,
And make poor England weep in streams of blood!
Let them not live to taste this land's increase,
That would with treason wound this fair land's peace!
Now civil wounds are stopp'd, peace lives again;
That she may long live here, God say—Amen!

King Richard III

CHAPTER 1
Organization

Very soon after the outbreak of the war for the Union, immediately, in fact, upon the commencement of actual operations in the field, it became painfully apparent that, however inferior the rank and file of the Confederate armies were in point of education and general intelligence to the men who composed the armies of the Union, however imperfect and rude their equipment and material, man for man they were the superiors of their northern antagonists in the use of arms. Recruited mainly from the rural districts (for the South had but few large cities from which to draw its fighting strength), their armies were composed mainly of men who had been trained to the skilful use of the rifle in that most perfect school, the field and forest, in the pursuit of the game so abundant in those sparsely settled districts. These men, who came to the field armed at first, to a large extent, with their favourite sporting or target rifles, and with a training acquired in such a school, were individually more than the equals of the men of the North, who were, with comparatively few exceptions, drawn from the farm, the workshop, the office or the counter, and whose life-long occupations had been such as to debar them from those pursuits in which the men of the South had gained their skill. Indeed, there were in many regiments in the northern armies men who had never even fired a gun of any description at the time of their enlistment.

On the other hand, there were known to be scattered throughout the loyal states, a great number of men who had made rifle shooting a study, and who, by practice on the target ground and at the country shooting matches, had gained a skill equal to that of the men of the South in any kind of shooting, and in long range practice a much greater degree of excellence.

There were many of these men in the ranks of the loyal army, but their skill was neutralized by the fact that the arms put into their hands, although the most perfect military weapons then known, were not of the description calculated to show the best results in the hands of expert marksmen.

Occasionally a musket would be found that was accurate in its shooting qualities, and occasionally such a gun would fall into the hands of a man competent to appreciate and utilize its best features. It was speedily found that such a gun, in the hands of such a man, was capable of results not possible to be obtained from a less accurate weapon in the hands of a less skilful man. To remedy this state of affairs, and to make certain that the best weapons procurable should be placed in the hands of the men best fitted to use them effectively, it was decided by the war department, early in the summer of 1861, that a regiment should be organized, to be called the First Regiment of United States Sharpshooters, and to consist of the best and most expert rifle shots in the Northern States. The detail of the recruiting and organization of this regiment was entrusted to Hiram Berdan, then a resident of the city of New York, himself an enthusiastic lover of rifle shooting, and an expert marksman.

Col. Berdan set himself earnestly at work to recruit and organize such a body of men as should, in the most perfect manner, illustrate the capacity for warlike purposes of his favourite weapon.

It was required that a recruit should possess a good moral character, a sound physical development, and in other respects come within the usual requirements of the army regulations; but, as the men were designed for an especial service, it was required of them that before enlistment they should justify their claim to be called "sharpshooters" by such a public exhibition of their skill as should fairly entitle them to the name, and warrant a reasonable expectation of usefulness in the field. To insure this it was ordered that no recruit be enlisted who could not, in a public trial, make a string of ten shots at a distance of two hundred yards, the aggregate measurement of which should not exceed fifty inches. In other words, it was required that the recruit should, in effect, be able to place ten bullets in succession within a ten-inch ring at a distance of two hundred yards.

Any style of rifle was allowed—telescopic sights, however, being disallowed—and the applicant was allowed to shoot from any position he chose, only being required to shoot from the shoulder.

Circular letters setting forth these conditions, and Col. Berdan authority, were issued to the governors of the loyal states, and, as a first result from the state of Vermont, Capt. Edmund Weston of Randolph applied for and received of Gov. Holbrook authority to recruit one company of sharpshooters, which was mustered into the service as Co. F, First United States Sharpshooters, and is the subject of this history.

Capt. Weston at once put himself in communication with well known riflemen in different parts of the state and appointed recruiting officers in various towns to receive applications and superintend the trials of skill, without which no person could be accepted.

The response was more hearty and more general than could have been expected, and many more recruits presented themselves than could be accepted—many of whom, however, failed to pass the ordeal of the public competition—and, as the event proved, more were accepted than could be legally mustered into the service.

All who were accepted, however, fully met the rigid requirements as to skill in the use of the rifle.

The company rendezvoused at Randolph early in September, 1861, and on the 13th of that month were mustered into the state service by Charles Dana. The organization of the company as perfected at this time was as follows:

Captain	Edmund Weston.
First Lieutenant	C. W. Seaton.
Second Lieutenant	M. V. B. Bronson.
First Sergeant	H. E. Kinsman.
Second Sergeant	E. W. Hindes.
Third Sergeant	Amos H. Bunker.
Fourth Sergeant	Milo C. Priest.
Fifth Sergeant	L. J. Allen.
First Corporal	Daniel Perry.
Second Corporal	Fred. E. Streeter.
Third Corporal	Ai Brown.
Fourth Corporal	W. C. Kent.
Fifth Corporal	H. J. Peck.
Sixth Corporal	W. H. Tafft.
Seventh Corporal	C. D. Merriman.
Eighth Corporal	C. W. Peck.
Bugler	Calvin Morse.
Wagoner	Edward F. Stevens.

Thus organized, the company, with one hundred and thirteen enlisted men, left the state on the same day on which they were mustered, and proceeded *via* New Haven and Long Island Sound to the rendezvous of the regiment at Weehawken Heights, near New York, where they went into camp with other companies of the regiment which had preceded them. On or about the 24th of September the regiment proceeded under orders from the war department to Washington, arriving at that city at a late hour on the night of the twenty-fifth, and were assigned quarters at the Soldiers' Rest, so well known to the troops who arrived at Washington at about that time. On the twenty-sixth they were ordered to a permanent camp of instruction well out in the country and near the residence and grounds of Mr. Corcoran, a wealthy resident of Washington of supposed secession proclivities, where they were for the first time in a regularly organized camp, and could begin to feel that they were fairly cut off at last from the customs and habits of civil life. Here they were regularly mustered into the service of the United States, thirteen enlisted men being rejected, however, to reduce the company to the regulation complement of one hundred enlisted men; so that of the one hundred and thirteen men charged to the company on the rolls of the Adjt. and Ins.-Gen. of Vermont, only one hundred took the field. Other companies from different states arrived at about the same time, and the regiment was at last complete, having its full complement of ten companies of one hundred men each.

The field and staff at this time was made up as follows:

Colonel	H. Berdan.
Lieutenant-Colonel	Frederick Mears.
Major	W. S. Rowland.
Adjutant	Floyd A. Willett.
Quarter-Master	W. H. Beebe.
Surgeon	G. C. Marshall.
Assistant Surgeon	Dr. Brennan.
Chaplain	Rev. Dr. Coit.

Only one of the field officers had had a military education or military experience. Lieutenant-Colonel Mears was an officer of the regular army, a thorough drill master and a strict disciplinarian. Under his efficient command the regiment soon began to show a marked and daily improvement that augured well for its future usefulness. The officers of the regimental staff were, each in his own department, able and painstaking men. The chaplain alone was not quite popular among the rank and file, and they rather envied the Second Regiment of Sharpshooters who were encamped near them, and whose chaplain, the Rev. Lorenzo Barber, was the *beau ideal* of an army chaplain. Tender hearted and kind, he was ever ready to help the weak and the suffering; now dressing a wound and now helping along a poor fellow, whose fingers were all thumbs and whose thoughts were too big for utterance (on paper), with his letter to the old mother at home; playing ball or running a foot race, beating the best marksmen at the targets, and finally preaching a rousing good sermon which was attentively listened to on Sunday. His *faith* was in the "*Sword of the Lord and of Gideon*," but his best *work* was put in with a twenty pound telescopic rifle which he used with wonderful effect. The original plan of armament contemplated the use exclusively of target or sporting rifles. The men had been encouraged to bring with them their favourite weapons, and had been told that the government would pay for such arms at the rate of sixty dollars each, while those who chose to rely upon the United States armouries for their rifles were to be furnished with the best implements procurable. The guns to be so furnished were to be breach loaders, to have telescopic sights, hair triggers, and all the requisites for the most perfect shooting that the most skilful marksman could desire.

Many of the men had, with this understanding, brought with them their own rifles, and with them target shooting became a pastime, and many matches between individuals and companies were made and many very short strings were recorded.

Under the stimulus of competition and organized practice great improvement was noted in marksmanship, even among those who had been considered almost perfect marksmen before. On one occasion President Lincoln, accompanied by Gen. Mc-Clellan, paid a visit to the camp and asked to be allowed to witness some of the sharpshooting of which he had heard so much.

A detail of the best men was made and a display of skill took place which, perhaps, was never before equalled. President Lincoln himself, as did Gen. McClellan, Col. Hudson and others of the staff, took part in the firing, the president using a rifle belonging to Corporal H. J. Peck of the Vermont company.

At the close of the exhibition Col. Berdan, being asked to illustrate the accuracy of his favourite rifle, fired three shots at different portions of the six hundred yard target; when having satisfied himself that he had the proper range, and that both himself and rifle could be depended upon, announced that at the next shot he would strike the right eye of the gaily coloured Zouave which, painted on the half of an A tent, did duty for a target at that range. Taking a long and careful aim, he fired, hitting the exact spot selected and announced beforehand. Whether partly accidental or not it was certainly a wonderful performance and placed Col. Berdan at once in the foremost rank of rifle experts. On the 28th of November, the day set apart by the governors of the loyal states as Thanksgiving Day, shooting was indulged by in different men of Co. F and other companies for a small prize offered by the field officers, the terms being two hundred yards, off hand, the shortest string of two shots to win. The prize was won from a large number of skilful contestants by Ai Brown of Co. F—his two shots measuring 4¼ inches, or each within 2⅛ inches of the centre.

On the 7th of December another regimental shooting match took place; the prize going this time to a Michigan man, his string of three shots, fired off hand at two hundred yards, measuring six inches. These records are introduced here simply for the purpose of showing the wonderful degree of skill possessed by these picked marksmen in the use of the rifle. But it was soon found that there were objections to the use in the field of the fine guns so effective on the target ground. The great weight of some of them was of itself almost prohibitory, for, to a soldier burdened with the weight of his knapsack, haversack and canteen, blanket and overcoat, the additional weight of a target rifle—many of which weighed fifteen pounds each, and some as much as thirty pounds—was too much to be easily borne.

It was also found difficult to provide the proper ammunition for such guns in the field, and finally, owing to the delicacy of the construction of the sights, hair triggers, etc., they were constantly liable to be out of order, and when thus disabled, of even less use than the smooth-bore musket, with buck and ball cartridge of fifty years before. Manufacturers of fine guns from all parts of our own country, and many from Europe, flocked to the camp of the sharpshooters offering their goods, each desirous of the credit of furnishing arms to a body of men so well calculated to use them effectively, and many fine models were offered. The choice of the men, however, seemed to be a modified military rifle made by the Sharpe Rifle Manufacturing Co., and a request was made to the war department for a supply of these arms. At this early day, however, the departments were full of men whose ideas and methods were those of a half a century gone by; and at the head of the ordinance department was a man who, in addition to being of this stamp, was the father of the muzzle loading Springfield rifle, then the recognized arm of the United States Infantry, and from him came the most strenuous opposition to the proposal to depart from the traditions of the regular army.

Gen. McClellan, and even the president himself, were approached on this subject, and both recognized the propriety of the proposed style of armament and the great capacity for efficient service possessed by the regiment when it should be once satisfactorily armed and fairly in front of the enemy. But the ordinance department was ever a block in the way; its head obstinately and stubbornly refusing to entertain any proposition other than to arm the regiment with the ordinary army musket; and, to add to the growing dissatisfaction among the men over the subject of arms, it became known that the promises made to them at the time of enlistment, that the government would pay them for their rifles at the rate of sixty dollars each, was unauthorized and would not be fulfilled; and also that the representations made to them with respect to telescopic breech loaders were likewise unauthorized. Discontent became general and demoralization began to show itself in an alarming form.

Some of the field officers were notoriously incompetent; the major, one of those military adventurers who floated to the surface during the early years of the war, particularly so; he was a kind of a modern Dalgetty without the courage or skill of his renowned prototype, rarely present in camp, and when there of little or no service. The Lieutenant-Colonel, a man of rare energy and skill in his profession, and whose painstaking care had made the regiment all that it was at that time, fearing the after effects of this demoralization on the efficiency of the command, and seeing opportunity for his talents in other fields, resigned; and on the 29th of November, 1861, Wm. Y. W. Ripley of Rutland, Vt., was appointed Lieutenant-Colonel, and Caspar Trepp, Captain of Co. A., was made Major. Lieutenant-Colonel Ripley had seen service only as Captain of Co. K, First Vermont Volunteers. Major Trepp had received a thorough military training in the army of his native Switzerland, and had seen active service in European wars. The regiment remained at camp of instruction under the immediate command of Lieut.-Col. Ripley, employed in the usual routine of camp duty, drills, etc., during the whole of the winter of 1861-62, particular attention being paid to the skirmish drill, in which the men became wonderfully proficient; and it is safe to say that for general excellence in drill, except the manual of arms, they were excelled by few volunteer regiments in the service. All orders were given by the sound of the bugle, and the whole regiment deployed as skirmishers could be as easily manoeuvred as a single company could be in line of battle. The bugle corps was under the charge of Calvin Morse of Co. F as chief bugler, and under his careful instruction attained to an unusual degree of excellence. All camp and other calls were sounded on the bugle, and the men found them pleasant little devices for translating curt and often rough English into music. They were bugled to breakfast and to dinner, bugled to guard mounting and bugled to battle, brigades moved and cavalry charged to the sound of the bugle. The men often found fanciful resemblances in the notes of the music to the words intended to be conveyed. Thus, the recall was sung as follows:

Come back again, come back again,
Come back, come back, come back again.

while the sick call was thus rendered into words:

Come to qui-nine, come to qui-nine,
Come to qui-i-nine, come to qui-i-nine.

They were not, on the whole, bad translations. The winter was an unusually severe one, and, as the enemy maintained a strict blockade of the Potomac, the supply of wood was often short, and some suffering was the result. The health of the regiment remained fairly good; measles, small pox, and other forms of camp diseases appeared, however, and Co. F, of course, suffered its share, losing by death from disease during the winter, Wm. T. Battles, Edward Fitz, Sumner E. Gardner and Geo. H. Johnson.

On the 20th of March, 1862, the regiment received orders to report to Major-Gen. Fitz John Porter, whose division then lay at Alexandria, Va., awaiting transportation to Fortress Monroe to join the army under McClellan. At this time the regiment was without arms of any kind, except for the few target rifles remaining in the hands of their owners, and a few old smooth bore muskets which had been used during the winter for guard duty. Shortly before this time the war department, perhaps wearied by constant importunity, perhaps recognizing the importance of the subject, had so far receded from its former position as to offer to arm the regiment with revolving rifles of the Colt pattern, and had sent the guns to the camp for issue to the men with promise of exchanging them for Sharpe's rifles at a later day. They were five chambered breech loaders, very pretty to look at, but upon examination and test they were found inaccurate and unreliable, prone to get out of order and even dangerous to the user. They were not satisfactory to the men, who knew what they wanted and were fully confident of their ability to use such guns as they had been led by repeated promises to expect, to good advantage. When, however, news came that the rebels had evacuated Manassas, and that the campaign was about to open in good earnest, they took up these toys, for after all they were hardly more, and turned their faces southward. Co. F was the first company in the regiment to receive their arms, and to the influence of their patriotic example the regiment owes its escape from what at one time appeared to be a most unfortunate embarrassment.

The march to Alexandria over Long Bridge was made in the midst of a pouring rain and through such a sea of mud as only Virginia can afford material for. It was the first time the regiment had ever broken camp, and its first hard march. It was long after dark when the command arrived near Cloud's mills; the headquarters of Gen. Porter could not be found, and it became necessary for the regiment to camp somewhere for the night. At a distance were seen the lights of a camp, which was found upon examination to be the winter quarters of the 69th New York in charge of a camp guard, the regiment having gone out in pursuit of the enemy beyond Manassas. A few persuasive words were spoken to the sergeant in command, and the tired and soaked sharpshooters turned into the tents of the absent Irishmen.

CHAPTER 2

The Peninsular Campaign

On the 22nd of March the regiment embarked on the steamer *Emperor*, bound for Fortress Monroe. The day was bright and glorious, the magnificent enthusiasm on every hand was contagious, and few who were partakers in that grand pageant will ever forget it. Alas, however, many thousands of that great army never returned from that fatal campaign. The orders required that each steamer, as she left her moorings, should pass up the river for a short distance, turn and pass down by Gen. Porter's flag-ship, saluting as she passed—a sort of military-naval review.

As the twenty-two steamers conveying this magnificent division thus passed in review, bands playing, colours flying and the men cheering, it was an inspiring spectacle for the young soldiers who were for the first time moving toward the enemy. The enthusiasm was kept up to fever heat until the leading steamers reached Mount Vernon, when, as though by order, the cheering ceased, flags were dropped to half-mast, the strains of *The Girl I Left Behind Me*, and *John Brown's Body*, gave way to funereal dirges, and all hats were doffed as the fleet passed the tomb of Washington. On the twenty-third the regiment disembarked at Hampton, Va., and went into camp at a point about midway between that place and Newport's News, where they remained several days, awaiting the arrival of the other divisions and the artillery and supplies necessary before the march on Yorktown could commence.

Hampton Roads was a scene of the greatest activity, hundreds of ships and steam transports lay at the docks discharging their cargoes of men and material, or at anchor in the broad waters adjacent awaiting their turn. Both army and navy here experienced a period of the most intense anxiety. Only a few days previous to the arrival of the first troops, the rebel iron-clad, *Merrimac*, had appeared before Newport's News, only a few miles away, and had made such a fearful display of her power for destruction as to excite the gravest apprehension lest she should again appear among the crowded shipping, sinking and destroying, by the simple battering power of her immense weight, these frail steamboats crowded with troops; but she had had a taste of the Monitor's quality, and did not apparently care to repeat the experiment. While thus awaiting the moment for the general advance, Fitz John Porter's division was ordered to make a reconnaissance in the direction of Great Bethel, the scene of the disaster of June 10, 1861. The division moved on two roads nearly parallel with each other. A body of sharpshooters led the advance of each column, that on the right being under the command of Lieut.-Col. Ripley, while those on the left were commanded by Col. Berdan.

This was the first time that the regiment had ever had the opportunity to measure its marching qualities with those of other troops; they had been most carefully and persistently drilled in this particular branch, and as they swept on, taking the full twenty-eight inch step and in regulation time, they soon left the remainder of the column far in rear, at which they were greatly elated, and when Capt. Auchmuty of Gen. Morell's staff rode up with the general's compliments and an inquiry as to "whether the sharpshooters intended to go on alone, or would they prefer to wait for support," their self-glorification was very great.

Later, however, they ceased to regard a march of ten or fifteen miles at their best pace as a joke. Co. F was with the right column, under Col. Ripley, and came for the first time under hostile fire. No serious fighting took place, although shots were frequently exchanged with the rebel cavalry, who fell back slowly before the Union advance. At Great Bethel a slight stand was made by the enemy, who were, however, soon dislodged by the steady and accurate fire of the sharpshooters, with some loss. Pushing on, the regiment advanced some three miles towards Yorktown, where, finding no considerable force of the enemy disposed to make a stand, and the object of the reconnaissance having been accomplished, both columns returned to camp near Fortress Monroe. The march had been a long and severe one for new troops, but Co. F came in without a straggler and in perfect order.

The experience of the day had taught them one lesson, however, and certain *gray* overcoats and Havelock hats of the same rebellious hue were promptly exchanged for others of a colour in which they were less apt to be shot by mistake by their own friends. The uniform of the regiment consisted of coats, blouses, pants and caps of green cloth; and leather leggings, buckling as high as the knee, were worn by officers and men alike. The knapsacks of the men were of the style then in use by the army of Prussia; they were of leather tanned with the hair on, and, although rather heavier than the regulation knapsack, fitted the back well, were roomy and were highly appreciated by the men. Each had strapped to its outside a small cooking kit which was found compact and useful. Thus equipped the regiment was distinctive in its uniform as well as in its service, and soon became well known in the army.

On the 3rd of April Gen. McClellan arrived at Fortress Monroe, and early on the morning of the fourth the whole army was put in motion toward Yorktown, where heavy works, strongly manned, were known to exist. The sharpshooters led the advance of the column on the road by which the Fifth Corps advanced, being that nearest the York River. Slight resistance was made by the enemy's cavalry at various points, but no casualties were experienced by Co. F on that day.

Cockeysville, a small hamlet some sixteen miles from Hampton, was reached, and the tired men of Co. F laid down in bivouac for the first time. Heretofore their camps, cheerless and devoid of home comforts as they sometimes were, had had some element of permanence; this was quite another thing, and what wonder if thoughts of home and home comforts flitted through their minds. Then, too, all supposed that on the morrow would occur a terrible battle (for the siege of Yorktown was not then anticipated); nothing less than immediate and desperate assault was contemplated by the men, and, as some complimentary remarks had been made to the regiment, and especial allusion to the effect those five shooting rifles, held in such trusty and skilful hands, would have in a charge, they felt that in the coming battle their place would be a hot and dangerous, as well as an honourable one. At daybreak on the morning of the fifth, in a soaking rain storm, the army resumed its march, the sharpshooters still in the advance, searching suspicious patches of woods, streaming out from the road to farm houses, hurrying over and around little knolls behind which danger might lurk, while now and then came the crack of rifles from a group across a field, telling of the presence of hostile cavalry watching the advance of the invaders. More strenuous resistance was met with than on the day before, but the rebels fell back steadily, if slowly. The rain fell continuously and the roads became difficult of passage for troops. The sharpshooters, however, fared better in this respect than troops of the line, for deployed as skirmishers, covering a large front, they could pick their way with comparative ease. At ten o'clock a.m., all resistance by rebel cavalry having ceased, the skirmishers emerged from dense woods and found themselves immediately in front of the heavy earth works before Yorktown. They were at once saluted by the enemy's artillery, and were now for the first time under the fire of shell.

Dashing forward one or two hundred yards, the skirmishers took position along and behind the crest of a slight elevation crowned by hedges and scattered clumps of bushes. The men of Co. F found themselves in a peach orchard surrounding a large farm house with its out-buildings. In and about these buildings, and along a fence running westwardly from the cluster of houses, Co. F formed its line, at a distance of some five hundred yards from a powerful line of breastworks running from the main fort in front of Yorktown to the low ground about the head of Warwick Creek.

Once in position, Co. F went at its work as steadily and coolly as veterans. Under the direction of a field officer, who watched the result with his glass, a few shots were fired by picked men at spots in the exterior slope of the works to ascertain the exact range, which was then announced and the order given, "Commence firing."

The rebels, ensconced in fancied security behind their strong works, and who up to that time had kept up a constant and heavy fire from their artillery, while their infantry lined the parapets, soon found reason to make themselves less conspicuous and to modify very essentially the tone of their remarks, which had been the reverse of complimentary. Gun after gun was silenced and abandoned, until within an hour every embrasure within a range of a thousand yards to the right and left was tenantless and silent. Their infantry, which at first responded with a vigorous fire, found that exposure of a head meant grave danger, if not death.

Occasionally a man would be found, who, carried away by his enthusiasm, would mount the parapet and with taunting cries seem to mock the Union marksmen, but no sooner would he appear than a score of rifles would be brought to bear, and he was fortunate indeed if he escaped with his life. At this point occurred the first casualty among the men of Co. F, Corp. C. W. Peck receiving a severe wound. During the day a small body of horsemen, apparently the staff and escort of a general officer, appeared passing from the village of Yorktown, behind the line of breastworks before spoken of, towards their right. When first observed little more than the heads of the riders were visible above the breastworks; near the western end of their line, however, the ground on which they were riding was higher, thus bringing them into plainer view, and as they reached this point every rifle was brought into use, and it appeared to observers that at least half the saddles in that little band were emptied before they could pass over the exposed fifty yards that lay between them and safety. While the sharpshooters had been successful in silencing the fire of the enemy's cannon, and almost entirely so that of their infantry, a few of the rebel marksmen, who occupied small rifle pits in advance of their line of works, kept up an annoying fire, from which the Union artillerists suffered severely.

These little strongholds had been constructed at leisure, were in carefully selected positions, usually behind a cover of natural or artificially planted bushes, and it was almost impossible to dislodge their occupants; every puff of smoke from one of them was, of course, the signal for a heavy fire of Union rifles on that spot; but sharpshooters who are worthy of the name will not continue long to fire at what they cannot see, and so, after one or two shots, the men would devote their attention to some other point, when the Confederate gunner, having remained quite at his ease behind his shelter, would peer out from behind his screen of bushes, select his mark, and renew his fire.

One spot was marked as the hiding place of a particularly obnoxious and skilful rifleman, and to him, Private Ide of Co. E of New Hampshire, who occupied a commanding position near the corner of an out house, devoted himself. Ide was one of the few men who still carried his telescopic target rifle. Several shots were exchanged between these men, and it began to take the form of a personal affair and was watched with the keenest interest by those not otherwise engaged, but fortune first smiled on the rebel, and Ide fell dead, shot through the forehead while in the act of raising his rifle to an aim. His fall was seen by the enemy, who raised a shout of exultation. It was short, however, for an officer, taking the loaded rifle from the dead man's hand, and watching his opportunity through the strong telescope, soon saw the triumphant rebel, made bold by his success, raise himself into view; it was a fatal exposure and he fell apparently dead.

At nine o'clock p.m. the sharpshooters were relieved by another regiment and retired to a point about half a mile in the rear, where the tired soldiers lay down after nearly twenty hours of continual marching and fighting. The fine position they had gained and held through the day, was regained, however, by the rebels by a night sally and was not reoccupied by the Union forces again for several days. On the next day, Gen. Porter, commanding the division, addressed the following highly complimentary letter to Col. Berdan:

Headquarters
Porter's Division
Third Army Corps
Camp near Yorktown
April 6, 1862
Col. Berdan
Commanding Sharpshooters
Colonel—The commanding general instructs me to say to you that he is glad to learn, from the admissions of the enemy themselves, that they begin to fear your sharpshooters. Your men have caused a number of the rebels to bite the dust. The commanding general is glad to find your corps are proving themselves so efficient, and trusts that this intelligence will encourage your men, give them, if possible, steadier hands and clearer eyes, so that when their trusty rifles are pointed at the foe, there will be one rebel less at every discharge. I am, colonel, very respectfully, your obedient servant,
Fred. T. Locke
A. A. G.

Gen. McClellan, believing the place too strong to be carried by assault, and his plans for turning the position having been disarranged by the detention in front of Washington of Gen. McDowell's corps, to which he had entrusted the movement, the army went into camp and settled down to the siege of Yorktown. The ensuing thirty days were full of excitement and danger, and Co. F had its full share. Several of the companies were detached and ordered to other portions of the army. Co. F, however, remained at regimental headquarters. Heavy details were made every day for service in the rifle pits, the men leaving camp and occupying their positions before daylight, and being relieved by details from other regiments after dark. Details were also frequently made for the purpose of digging advanced rifle pits during the night. These pits were approached by zigzags, and could only be reached during the hours of daylight by crawling on the hands and knees, and then only under circumstances of great danger. They were pushed so far to the front that, when the evacuation took place on the night of the 3rd of May, they were hardly more than one hundred yards from the main rebel line of works, and hardly half as far from the rebel rifle pits. Frequent sharp conflicts took place between bodies of rebel and Union soldiers striving for the same position on which to dig a new rifle pit, in several of which Co. F took a prominent part and suffered some loss.

So close were the opposing lines at some places that sharp-shooting became almost impossible for either side, as the exposure of so much as a hand meant a certain wound.

In this state of affairs the men would improvise loop holes by forcing sharpened stakes through the bank of earth in front of the pits, through which they would thrust the barrels of their breach loaders, over which they would keenly watch for a chance for a shot, and woe to that unfortunate rebel who exposed even a small portion of his figure within the circumscribed range of their vision.

The regimental camp before Yorktown was beautifully situated near the York River and not far from army headquarters. Great rivalry existed between the different companies as to which company street should present the neatest appearance, and the camp was very attractive to visitors and others. The officers mess was open to all comers and was a constant scene of visiting and feasting. For a few days, it is true, the troops, officers and men alike, were on short rations, but as soon as the river was opened and docks constructed, the necessities, and even the luxuries of life were abundant. At this camp the first instalment of the much desired and long promised Sharpe rifles arrived. Only one hundred were received in the first consignment, and they were at once issued to Co. F as an evidence of the high esteem in which that company was held by the officers of the regiment, and as a recognition of its particularly good conduct on several occasions—it was a compliment well deserved. On the night of the 3rd of May, the rebels kept up a tremendous fire during the whole night. Heavy explosions, not of artillery, were frequent, and it was evident that some move of importance was in progress. At an early hour the usual detail of sharpshooters relieved the infantry pickets in the advanced rifle pits, and soon after daylight it became apparent to them that matters at the front had undergone a change, and cautiously advancing from their lines they found the rebel works evacuated.

Pressing forward over the earth works which had so long barred the way, the sharpshooters were the first troops to occupy the village of Yorktown, where they hauled down the garrison flag which had been left flying by the retreating rebels. All was now joyous excitement; what was considered a great victory had been gained without any considerable loss of life—a consideration very grateful for the soldier to contemplate. Seventy-two heavy guns were abandoned by the rebels, which, though of little use to them, and of less to us, by reason of their antiquated styles, were still trophies, and so, valuable.

Regimental and brigade bands, which, together with drum and bugle corps, had been silent for a month, by general orders (for the rebels had kept up a tremendous fire on every thing they saw, heard or suspected), now filled the air with many a stirring and patriotic strain. Salutes were fired, and with the balloon, used for observing the movements of the enemy, floating in the air overhead, one could easily believe himself to be enjoying a festival, and for a moment forget the miseries of war. At York-town the rest of the regiment received their Sharpe's rifles and, with the exception of a few men who still clung to their muzzle loaders, the command was armed with rifles of uniform calibre, and which were entirely satisfactory to those who bore them. The Colt's five shooters were turned in without regret; for, although they had done fairly good service, they were not quite worthy of the men in whose hands they were placed.

On the 5th of May was fought the battle of Williamsburgh, on which hard fought field two companies of the regiment, A and C, bore an honourable part—Co. F, however, was with the part of the command retained in front of Yorktown. The guns were plainly heard at the camp, and painful rumours began to be circulated. At about ten o'clock a.m. there came an order to prepare to march at once, with three day's cooked rations; the concluding words of the brief written message, "prepare for hard fighting," were full of significance, but they were received with cheers by the men who were tired of rifle pit work, and desired ardently an opportunity to measure their skill with that of the boasted southern riflemen in the field—a desire that was shortly to be gratified to an extent satisfactory to the most pronounced glutton among them. The preparations were soon made, and the regiment formed on the colour line, but the day passed and the order to march did not come. The battle of Williamsburgh was over. On the evening of the eighth the regiment was embarked on the steam transport *State of Maine*, and under convoy of the gun boats proceeded up the York River to West Point where they disembarked on the afternoon of the ninth, finding the men of Franklin's division, which had preceded them, in position. Franklin's men had had a sharp fight the day before with the rear guard of the Confederate army, but were too late to cut off the retreat of the main body, whose march from the bloody field of Williamsburgh had been made with all the vigour that fear and necessity could inspire. Here the sharpshooters remained in bivouac until the thirteenth, when they were put in motion again towards Richmond. The weather was warm, the roads narrow and dusty, water scarce and the march a wearisome one. Rumours of probable fighting in store for them at a point not far distant were rife, but no enemy was found in their path on that day, and near sundown they went into camp at Cumberland Landing on the Pamunkey.

On the fourteenth the regiment was reviewed by Secretary Seward, who made a short visit to the army at this time. On the fifteenth they marched to White House, a heavy rain storm prevailing through the entire day. The sharpshooters were in support of the cavalry and had in their rear a battery, the guns of which were frequently stalled in the deep mud, out of which they had often to be lifted and pulled by sheer force of human muscle. The march was most fatiguing, and although commenced at half-past six a.m., and terminating at four p.m., only about six miles were gained. White House was a place of historic interest, since it was here that Washington wooed and married his wife; a strict guard was kept over it and its surroundings, and it was left as unspoiled as it was found. Above White House the river was no longer navigable, and the York River railroad, which connects Richmond, some twenty miles distant, with the Pamunkey at this point, was to be the future line of supply for the army. On the nineteenth the troops again advanced, camping at Turnstall's Station that night and at Barker's Mill on the night of the twentieth. On the twenty-sixth they passed Cold Harbor, a spot on which they were destined to lose many good and true men two years later, and went into camp near the house of Dr. Gaines, and were now fairly before Richmond, the spires of which could be seen from the high ground near the camp. On the morning of the twenty-seventh, at a very early hour, there came to regimental headquarters an order couched in words which had become familiar:

This division will march at daylight in the following order: First, the sharpshooters.

Three days cooked rations and one hundred rounds of ammunition were also specified. This looked like business, and the camp became at once a scene of busy activity. At the appointed hour, in the midst of a heavy rain shower, the column was put in march, but not, as had been anticipated, towards the enemy who blocked the road to the rebel capital. The line of march was to the northward towards Hanover Court House.

As the head of the column approached the junction of the roads leading respectively to Hanover Court House and Ashland, considerable resistance was met with from bodies of rebel cavalry supported by a few pieces of light artillery and a small force of infantry. At the forks of the road a portion of Branch's brigade of North Carolina troops were found in a strong position, prepared to dispute the passage. This force were soon dislodged by the sharpshooters, the Twenty-Fifth New York, a detachment from a Pennsylvania regiment and Benson's battery, and retreated in the direction of Hanover Court House. Prompt pursuit was made and many prisoners taken, together with two guns. Martindale's brigade was left at the forks of the road before spoken of, to guard against an attack on the rear from the direction of Richmond, while the rest of the division pushed on to destroy, if possible, the bridges at the points where the Richmond & Fredericksburgh and the Virginia Central railroads cross the North and the South Anna Rivers; the destruction of these bridges being the main object of the expedition, although it was hoped and expected that the movement might result in a junction of the forces under McDowell, then at Fredericksburgh only forty miles distant from the point to which Porter's advance reached, with the right of McClellan's army, when the speedy fall of Richmond might be confidently expected.

The sharpshooters accompanied the column which was charged with this duty. The cavalry reached the rivers and succeeded in completing the destruction of the bridges, when ominous reports began to come up from the rear, of heavy forces of the enemy having appeared between this isolated command and the rest of the army twenty miles to the southward. Firing was heard distinctly, scattering and uncertain at first, but soon swelling into a roar that gave assurance of a hotly contested engagement.

The column was instantly faced about, not even taking time to counter-march, and taking the double quick—left in front—made all haste to reach the scene of the conflict. The natural desire to help their hard pressed comrades was supplemented by a conviction that their own safety could only be secured by a speedy destruction of the force between them and their camp, and the four or five miles of road, heavy with mud, for, as usual, the rain was falling fast, were rapidly passed over. As they neared the field of battle the sharpshooters, who had gained what was now the head of the column, were rapidly deployed and with ringing cheers passed through the ranks of the 2nd Maine, opened for the purpose, and plunged into the woods where the enemy were posted. The spirit of the rebel attack was already broken by the severe losses inflicted upon them by Martindale's gallant brigade which, although out-numbered two to one, had clung desperately to their all important position; and when the enemy heard the shouts of this relieving column, and caught sight of their advancing lines, a panic seized them and they fled precipitately from the field. Pursuit was made and many prisoners taken, who, with those captured in the earlier part of the day, swelled the total to over seven hundred. Two guns were also taken, in the capture of which Co. F bore a prominent part. This affair cost the Union forces four hundred men; the loss, however, principally falling on Martindale's brigade, who bore the brunt of the rear attack. The sharpshooters lost only about twenty men, killed and wounded—three of whom, Sergt. Lewis J. Allen, Benjamin Billings and W. F. Dawson were of Co. F; Dawson died on the 1st of June from the effects of his wound.

The regiment, however, met with a great loss on that day by the capture of its surgeon, Guy C. Marshall, who, with other surgeons and attendants, was surprised by a sudden attack on the field hospital by the enemy's cavalry. Dr. Marshall never rejoined the regiment. Being sent to Libby Prison, he was, with other surgeons, allowed certain liberties in order that he might be the more useful in his professional capacity. Placed upon his parole he was allowed, under certain restrictions, to pass the prison guards at will, for the purpose of securing medicines, etc., for use among the sick prisoners. The terrible sufferings of his comrades, caused mainly by what he believed to be intentional neglect, aroused all the sympathy of his tender nature, and as the days passed and no attention was paid to his protests or efforts to get relief, his intense indignation was aroused. Taking advantage of his liberty to pass the guards, he succeeded in getting an audience with Jefferson Davis himself. It is probable that his earnestness led him into expressions of condemnation too strong to be relished by the so called President. Howsoever it was, his liberty was stopped and he was made a close prisoner. He continued his labours, however, with such scanty means as he could obtain until, worn out by his over exertions, and with his great heart broken by the sight of the suffering he was so powerless to relieve, he died,—as truly the death of a hero as though he had fallen at the head of some gallant charge in the field. He was a true man, and those who knew him best will always have a warm and tender remembrance of him.

On the twenty-ninth, the whole command returned to their camp near Gaines Hill. The experience of Co. F for the next thirty days was similar to that of Yorktown—daily details for picket duty were made, and always where the danger was greatest; for, as it was the province of the sharpshooter to shoot some body, it was necessary that he should be placed where there was some one to shoot. In a case of this kind, however, one cannot expect to give blows without receiving them in return, hence it came about that the sharpshooters were constantly in the most dangerous places on the picket line. At some point in the Union front, perhaps miles away, it would be found that a few rebel sharpshooters had planted themselves in a position from which they gave serious annoyance to the working parties and sometimes inflicted serious loss, and from which they could not readily be dislodged by the imperfect weapons of the infantry. In such cases calls would be made for a detail of sharpshooters, who would be gone sometimes for several days before returning to camp, always, however, being successful in removing the trouble.

On the thirty-first, the guns of Fair Oaks were distinctly heard, and early the next morning the Fifth Corps, to which the regiment was now attached, was massed near the head of New Bridge on the Chicahominy, with the intention of forcing a passage at this place to try to convert the repulse of the rebels at Fair Oaks on the day before into a great disaster. The swollen condition of the river, however, which had proved so nearly fatal to the Union forces on the day of Fair Oaks, became now the safety of the rebels. A strong detachment of the sharpshooters, including some men from Co. F, were thrown across the river at New Bridge to ascertain whether the water covering the road beyond was fordable for infantry. This detachment crossed the bridge and passed some distance along the road, but finding it impracticable for men, so reported and the attempt was abandoned.

No incidents of unusual interest occurred to the Vermonters after June 1st until the movements commenced which culminated in what is known in history as the seven days battle, commencing on the 25th of June at a point on the right bank of the Chicahominy at Oak Grove, and ending on the first of July at Malvern Hill on the James River.

For some days rumours of an unfavourable nature had been circulating among the camps before Richmond, of disasters to the Union forces in the valley. It was known that Stonewall Jackson had gone northward with his command, and that he had appeared at several points in northern Virginia under such circumstances and at such times and places as caused serious alarm to the government at Washington for the safety of the capital. To the Army of the Potomac, however, it seemed incredible that so small a force as Jackson's could be a serious menace to that city, and preparations for a forward movement and a great and decisive battle went steadily on. On the 25th of June, Hooker advanced his lines near Oak Grove, and after severe fighting forced the enemy from their position which he proceeded to fortify, and which he held. On the night of that day, the army was full of joyous anticipation of a great victory to be gained before Jackson could return from his foray to the north. On the morning of the twenty-sixth, however, scouts reported Jackson, reinforced by Whiting's division, at Hanover Court House pressing rapidly forward, with 30,000 men, toward our exposed right and rear. At the same time large bodies of the enemy were observed crossing the Chicahominy at Meadow Bridge, above Mechanicsville. It was at once apparent that the Army of the Potomac must abandon its advance on Richmond, for the time at least, and stand on its defence. McCall, with his division of Pennsylvania reserves, occupied a strong position on the left bank of Beaver Dam creek, a small affluent of the Chicahominy, near Mechanicsville, about four miles north of Gaines Hill, and this command constituted the extreme right of the Union army. On this isolated body it was evident that the first rebel attack would fall.

At about three o'clock p.m. the division of the rebel General A. P. Hill appeared in front of McCall's line, and severe fighting at once commenced. About one hour later Branch's division arrived to the support of the rebel general, and vigorous and repeated assaults were made at various points on the Union line; the fighting at Ellison's Mills being of a particularly desperate character. Porter's old division, now commanded by Morell, was ordered up from its camp at Gaines Hill to the assistance of the troops so heavily pressed at Mechanicsville. The sharpshooters, being among the regiments thus detailed, left their tents standing, and in light marching order, and with no rations, moved out at the head of the column. Arriving at the front they took post in the left of the road, in the rear of a rifle pit occupied by a battalion of Pennsylvania troops and on the right of a redoubt in which was a battery of guns. It was now nearly dark, the force of the attack was spent, and the sharpshooters had but small share of the fighting. The night was spent in this position, and the rest of the soldiers was unbroken, except by the cries and moans of the rebel wounded, many of whom lay uncared for within a few yards of the Union line. Some of the men of Co. F, moved by pity for the sufferings of their enemies, left their lines to give them assistance; they were fired on, however, by the less merciful rebels and had to abandon the attempt. Before daylight the order was whispered down the line to withdraw as silently as possible. The men were especially cautioned against allowing their tin cups to rattle against their rifles, as the first sign was sure to be the signal for a rebel volley. Cautiously the men stole away, and, as daylight appeared, found themselves alone.

They were the rear guard and thus covered the retreat of the main body to Gaines Hill. As they approached the camp they had left on the preceding afternoon a scene of desolation and destruction met their astonished eyes. Enormous piles of quartermaster and commissary stores were being fired, tents were struck, the regimental baggage gone, and large droves of cattle were being hurried forward towards the lower bridges of the Chicahominy—the retreat to the James had commenced. Halting for a few minutes amidst the ruins of their abandoned camp where, however, they found the faithful quartermaster-sergeant with a scanty supply of rations, very grateful to men who had eaten nothing for twenty hours and expected nothing for some time to come. They hastily commenced the preparation of such a modest breakfast as was possible under the circumstances, but before it could be eaten the pursuing rebels were upon them, and the march towards the rear was resumed. A mile further and they found the Fifth Corps, which was all there was of the army on the south bank of the Chicahominy, in line of battle prepared to resist the attack of the enemy, which it was apparent to all would be in heavy force. The position was a strong one, and the little force—small in comparison to that which now appeared confronting it—were disposed with consummate skill. Dust—for the day was intensely hot and dry—arising in dense clouds high above the tree tops, plainly denoted the line of march, and the positions of the different rebel columns as they arrived on the field and took their places in line of battle.

Deserters, prisoners, and scouts, all agreed that Jackson, who had not been up in time to take part in the battle of the previous day as had been expected, was now at hand with a large force of fresh troops, and it was apparent that the Fifth Corps was about to become engaged with nearly the whole of the rebel army. Any one of three things could now happen, as might be decided by the Union commander. The force on Gaines Hill might be re-enforced by means of the few, but sufficient, bridges over the Chicahominy and accept battle on something like equal terms; or the main army on the right bank of the river might take advantage of the opportunity offered to break through the lines in its front, weakened as they must be by the absence of the immense numbers detached to crush Porter on the left bank; or the Fifth Corps might by a great effort, unassisted, hold Lee's army in check long enough to enable the Union army to commence in an orderly manner its retreat to the James. Whichever course might be decided upon, it was evident that this portion of the army was on the eve of a desperate struggle against overwhelming odds, and each man prepared himself accordingly.

In front of Morell's division, to which the sharpshooters were attached, was a deep ravine heavily wooded on its sides, and through which ran a small stream, its direction being generally northeast, until it emptied into the Chicahominy near Woodbury's Bridge. The bottom of the ravine was marshy and somewhat difficult of passage, and near the river widened out and took the name of Boatswain's Swamp. On the far side of this ravine the sharpshooters were deployed to observe the approach of the enemy and to receive their first attack. In their front the ground was comparatively open, though somewhat broken, for a considerable distance. At half-past two p.m. the enemy's skirmishers appeared in the rolling open country, and desultory firing at long range commenced. Soon, however, the pressure became more severe, and a regiment on the right of the sharpshooters having given way, they, in their turn, were forced slowly back across the marshy ravine and part way up the opposite slope; here, being re-enforced, they turned on and drove the rebels back and reoccupied the ground on which they first formed, soon, however, to be forced back again. So heavily had each of the opposing lines been supported that the affair lost its character as a picket fight, and partook of the nature of line of battle fighting. The troops opposed at this time were those of A. P. Hill, who finally, by sheer weight of numbers, dislodged the sharpshooters and their supports from the woods and permanently held them. They were unable, however, to ascend the slope on the other side, and the main federal line was intact at all points. There was now an interval of some half an hour, during which time the infantry were idle; the artillery firing, however, from the Union batteries on the crest of the hill was incessant, and was as vigorously responded to by the rebels. From the right bank of the Chicahominy a battery of twenty pound Parrots, near Gen. W. F. Smith's headquarters, was skilfully directed against the rebel right near and in front of Dr. Gaines' house. At six o'clock p.m. Slocum's division of Franklin's corps was ordered across to the support of Porter's endangered command.

At seven o'clock the divisions of Hill, Longstreet, Whiting and Jackson were massed for a final attack on the small but un-dismayed federal force, who yet held every inch of the ground so desperately fought for during five long hours.

Whiting's division led the rebel assault with Hood's Texan brigade in the front line. The attack struck the centre of the line held by Morell's division, and so desperate was the assault and so heavily supported, that Morell's tired men were finally forced by sheer weight of masses to abandon the line which they had so long and so gallantly held. Had the rebels them-selves been in a position to promptly pursue their advantage, the situation would have been most perilous to the Union forces. The enemy had now gained the crest of the hill which commanded the ground to the rear as far as the banks of the Chicahominy. This deep and treacherous stream, crossed but by few bridges—and they, with one exception, at a considerable distance from the field of battle—offered an effectual barrier to the passage of the routed army.

But while the federals had suffered severely, the losses of the rebels had been far greater. The disorganization and demor-alization among the victors was even greater than among the vanquished; and before they could reform for further advance the beaten federals had rallied on the low ground nearer the river and formed a new line which, in the gathering darkness, undoubtedly looked to the rebels, made cautious by experi-ence, more formidable than it was in fact. Their cavalry ap-peared in great force on the brow of the hill, but the expected charge did not come; they had had fighting enough and rested content with what they had gained. The least desirable of the three choices offered to the Union commander had been taken, as it appeared, but a precious day had been gained to the army already in its retreat to the James. A fearful price had been paid for it, however, by the devoted band who stood between that retreating army and the flushed and victorious enemy. Of the eighteen thousand men who stood in line of battle at noon, only twelve thousand answered to the roll call at night. One-third of the whole, or six thousand men, had fallen. They had done all that it was possible for men to do, and only yielded to superior numbers. It is now known that less than 25,000 men were left for the defence of Richmond; the rest of the rebel forces, or over 55,000 men, had been hurled against this wing of the Union army hoping to crush it utterly, and the attempt had failed.

Co. F had done its full share in the work of the day, and, although out of ammunition, retained its position with other companies of the regiment on the front line until the general disruption on the right and left compelled their retirement from the field. Tired, hungry and disheartened, they lay down for the night on the low ground a mile or more in the rear for a few hours of repose. At about eleven o'clock p.m. they were aroused and put in motion, crossing the Chicahominy at Woodbury's Bridge and going again into bivouac on the high ground near the Trent Hospital some distance in the rear of the ground held by the Vermont brigade on the northern, or right, bank of the river. During the night the entire corps was withdrawn and the bridges destroyed. A fresh supply of ammunition was obtained and issued at daylight, and at ten o'clock a.m. the sharpshooters, with full cartridge boxes, but empty haversacks, took up their line of march towards the James. In this action the regiment lost heavily in killed and wounded. B. W. Jordan and Jas. A. Read of Co. F were mortally wounded, and E. H. Himes severely wounded. Passing Savage Station, where the 5th Vermont suffered so severely on the next day, the regiment crossed White Oak Swamp before dark on the twenty-eighth, and went into bivouac near the head of the bridge.

Wild rumours of heavy bodies of Confederate troops, crossing the Chicahominy at points lower down prepared to fall upon the exposed flank and rear of the federals were prevalent, and the dreaded form of Stonewall Jackson seemed to start from every bush.

During the night, which was intensely dark, the horses attached to a battery got loose by some means and, dashing through a portion of the ground occupied by other troops, seemed, with their rattling harness, to be a host of rebel cavalry. A bugle at some distance sounded the assembly, drums beat the long roll, and in the confusion of that night alarm it seemed as though a general panic had seized upon all. The sharpshooters, like all others, were thrown into confusion and momentarily lost their sense of discipline and disappeared. When the commanding officer, perhaps the last to awake, came to look for his command they were not to be found; with the exception of Calvin Morse, bugler of Co. F, he was alone. The panic among the sharpshooters, however, was only momentary; the first blast of the well known bugle recalled them to a sense of duty, and, a rallying point being established, the whole command at once returned to the line reassured and prepared for any emergency.

At daylight the march was resumed and continued as far as Charles City cross roads, or Glendale, the junction of two important roads leading from Richmond south-easterly towards Malvern Hill; the lower, or Newmarket road, being the only one by which a rebel force moving from the city could hope to interpose between the retreating federals and the James.

The sharpshooters were thrown out on this road some two miles with instructions to delay as long as possible the advance of any body of the enemy who might approach by that route. This was the fourth day for Co. F of continuous marching and fighting; they had started with almost empty haversacks, and it had not been possible to supply them. The country was bare of provisions, except now and then a hog that had so far escaped the foragers. A few of these fell victims to the hunger of the half-starved men; but, with no bread or salt, it hardly served a better purpose than merely to sustain life. To add to their discomforts the only water procurable was that from a well near by which was said to have been poisoned by the flying owner of the plantation; his absence, with that of every living thing upon the place, made it impossible to apply the usual and proper test, that of compelling the suspected parties to, themselves, drink heartily of the water. A guard was therefore placed over the well, and the thirsty soldiers were compelled to endure their tortures as best they could. The day passed in comparative quiet; only a few small bodies of rebel cavalry appeared to contest the possession of the road, and they being easily repulsed. Late in the afternoon the sharpshooters were recalled to the junction of the roads, where they rested for a short time to allow the passage of another column. At this point a single box of hard bread was procured from the cook in charge of a wagon conveying the mess kit of the officers of a battery; this was the only issue of rations made to the regiment from the morning of the 25th of June until they arrived at Harrison's Landing on the 2nd of July, and, inadequate as it was, it was a welcome addition to their meagre fare.

At dark the regiment marched southwardly on a country road narrow and difficult, often appearing no more than a path through the dense swamp; the night, intensely dark, was made more so by the gloom of the forest, and all night the weary unfed men toiled along. At midnight the column was halted for some cause, and while thus halted another of those unaccountable panics took place—in fact, in the excited condition of the men, enfeebled by long continued labours without food, a small matter was sufficient to throw them off their balance; and yet these very men a few hours later, with an enemy in front whom they could see and at whom they could deal blows as well as receive them, fought and won the great battle of Malvern Hill. During the night Co. F. with one or two other companies were detailed to accompany Gen. Porter and others on a reconnaissance of the country to the left of the road on which the column was halted. With a small force in advance as skirmishers, they passed over some two miles of difficult country, doubly so in the darkness of the night, striking and drawing the fire of the rebel pickets. This being apparently the object of the movement, the skirmishers were withdrawn and the command rejoined the main column. So worn and weary were they that whenever halted even for a moment, many men would fall instantly into a sleep from which it would require the most vigorous efforts to arouse them. Shortly before daylight they were halted and allowed to sleep for an hour or two, when, with tired and aching bodies, they continued their march. At noon they passed over the crest of Malvern Hill and before them lay, quiet and beautiful in the sunlight, the valley of the James; and, at the distance of some three miles, the river itself with Union gun boats at anchor on its bosom.

It was a welcome sight to those who had been for six long days marching by night and fighting by day. It meant, as they fondly believed, food and rest, and they greeted the lovely view with cheers of exultation. But there were further labours and greater dangers in store for them before the longed for rest could be obtained. Passing over the level plateau known as Malvern Hill, they descended to the valley and went into bivouac. Here was at least water, and some food was obtained from the negroes who remained about the place.

No sooner were ranks broken and knapsacks unslung than the tired and dirty soldiers flocked to the banks of the beautiful river, and the water was soon filled with the bathers, who enjoyed this unusual luxury with keen relish.

The bivouac of the regiment was in the midst of a field of oats but recently cut and bound, and the men proceeded to arrange for themselves couches which for comfort and luxury they had not seen the like of since they left the feather beds of their New England homes. Their repose, even here, was, however, destined to be of short duration; for hardly had they settled themselves for their rest when the bugles sounded the general, and the head of the column, strangely enough, turned northward. Up the steep hill, back over the very road down which they had just marched, they toiled, but without murmur or discontent, for *this* movement was *towards* the rebels, and not away from them. Inspiring rumours began to be heard; where they came from, or how, no one knew, but it was said that McCall and Sumner had fought a great battle on the previous day, that the rebel army was routed, that Lee was a prisoner, that McClellan was in Richmond, and the long and short of it was that the Union army had nothing more to do but to march back, make a triumphal entry into the captured stronghold, assist at that often anticipated ceremony which was to consign "*Jeff. Davis to a sour apple tree,*" be mustered out, get their pay and go home. When they arrived on the plateau, however, a scene met their eyes that effectually drove such anticipations from their minds. A mile away, just emerging from the cover of the forest, appeared the forms of a number of men; were they friends or enemies? Glasses were unslung and they were at once discovered to be federals. Momentarily their numbers increased, and soon the whole plain was covered with blue coated troops, but they were without order or organization, many without arms, and their faces bearing not the light of successful battle, but dull with the chagrin of defeat. The story was soon told. Sumner and McCall had fought a battle at Charles City crossroads, but had been forced to abandon the field with heavy loss in men and guns. Instead of a triumphant march to Richmond, the Fifth Corps was again to interpose between the flushed and confident rebels and the retreating federals—but not, as at Gaines Hill, alone. This was late in the afternoon of June 30th. That night the sharpshooters spent in bivouac near the ground on which they were to fight the next day. At dawn on the 1st of July the men were aroused, and proceeding to the front were ordered into line as skirmishers, their line covering the extreme left of the Union army directly in front of the main approach to the position. Malvern Hill, so called, is a hill only as it is viewed from the southern or western side; to the north and east the ground is only slightly descending from the highest elevation. On the western side, flowing in a southerly direction, is a small stream called Turkey Run, the bed of the stream being some one hundred feet lower than the plateau. On the south, toward the James, the descent is more precipitous. The approaches were, as has been stated, from the north where the ground was comparatively level and sufficiently open to admit of rapid and regular manoeuvres. The position taken by the Union army was not one of extraordinary strength, except that its flanks were well protected by natural features: its front was but little higher than the ground over which the enemy must pass to the attack, and was unprotected by natural or artificial obstacles. No earth works or other defences were constructed; although the "*lofty hill, crowned by formidable works,*" has often figured in descriptions of this battle. The simple truth is it was an open field fight, hotly contested and gallantly won.

The Union artillery, some three hundred guns, was posted in advantageous positions, some of the batteries occupying slight elevations from which they could fire over the heads of troops in their front, the most of them, however, being formed on the level ground in the intervals between regiments and brigades. The gun boats were stationed in the river some two miles distant, so as to cover and support the left flank, and it was expected that great assistance would be afforded by the fire of their immense guns.

Porter's corps held the extreme left, with its left flank on Turkey Run, Morell's division forming the front line with headquarters at Crew's house. Sykes' division, composed mostly of regulars, was in the second line. McCall's division was held in reserve in rear of the left flank. On the right of Morell's line thus formed, came Couch's division; further to the right the line was refused, and the extreme right flank rested on the James; but with this portion of the line we have little to do. The main attack fell on the Fifth Corps, involving to some extent Couch's troops next on the right. In this order the army awaited the onset. In front of Morell's division stretched away a field about half a mile in length, bounded at its opposite extremity by heavy woods.

Nearly level in its general features, there extended across it at a distance of about one-third of a mile from the federal front, and parallel with it, a deep ravine, its western end debouching into the valley formed by Turkey Run. This open field was covered at this time with wheat just ready for the harvest.

Along the north side of this ravine, covered from view by the waving wheat, the sharpshooters were deployed at an early hour and patiently awaited the attack of the enemy. A few scattered trees afforded a scanty supply of half grown apples which were eagerly seized upon by the famished men, who boiled them in their tin cups and thus made them fairly palatable; by such poor means assuaging as best they could the pangs of hunger.

At about twelve o'clock heavy clouds of dust arising in the north announced the approach of the Confederate columns, and soon after scouts and skirmishers began to make their presence known by shots from the edge of the woods, some two hundred yards distant, directed at every exposed head. A puff of smoke from that direction, however, was certain to be answered by a dozen well aimed rifles from the sharpshooters, and the rebel scouts soon tired of that amusement. In the meantime the artillery firing had become very heavy on both sides, our own depressing their muzzles so as to sweep the woods in front; the effect of this was to bring the line of fire unpleasantly near the heads of the advanced sharpshooters. The gun boats also joined in the cannonade, and as their shells often burst short, over and even behind the line of skirmishers, the position soon became one of grave danger from both sides.

At about half-past two the artillery fire from the rebel line slackened perceptibly, and soon appeared, bursting from the edge of the forest, a heavy line of skirmishers who advanced at a run, apparently unaware of any considerable force in their front. Bugler Morse of Co. F, who accompanied the commanding officer as chief bugler on that day, was at once ordered to sound commence firing, and the sharpshooters sent across the field and into the lines of the oncoming rebels, such a storm of lead from their breach loading rifles as soon checked their advance and sent them back to the cover of the woods in great confusion and with serious loss. The repulse was but momentary, however,

45

for soon another line appeared so heavily re-enforced that it was more like a line of battle than a skirmish line. Still, however, the sharpshooters clung to their ground, firing rapidly and with precision, as the thinned ranks of the Confederates, as they pressed on, attested. They would not, however, be denied, but still came on at the run, firing as they came. At this moment the sharpshooters became aware of a force of rebel skirmishers on their right flank, who commenced firing steadily, and at almost point blank range, from the shelter of a roadway bordered by hedges. The bugle now sounded retreat, and the sharpshooters fell back far enough to escape the effect of the flank fire when they were halted and once more turned their faces to the enemy. The tables were now turned; the rebels had gained the shelter of the ravine, and were firing with great deliberation at our men who were fully exposed in the open field in front of the Crew house. Still the sharpshooters held their ground, and, by the greater accuracy of their fire, combined with the advantage of greater rapidity given by breach loaders over muzzle loaders, kept the rebels well under cover. Having thus cleared the way, as they supposed, for their artillery, the rebels sought to plant a battery in the open ground on the hither side of the woods which had screened their advance. The noise of chopping had been plainly heard for some time as their pioneers laboured in the woods opening a passage for the guns. Suddenly there burst out of the dense foliage four magnificent stray horses, and behind them, whirled along like a child's toy, the gun. Another and another followed, sweeping out into the plain. As the head of the column turned to the right to go into battery, every rifle within range was brought to bear, and horses and men began to fall rapidly. Still they pressed on, and when there were no longer horses to haul the guns, the gunners sought to put their pieces into battery by hand; nothing, however, could stand before that terrible storm of lead, and after ten minutes of gallant effort the few survivors, leaving their guns in the open field, took shelter in the friendly woods. Not a gun was placed in position or fired from that quarter during the day. This battery was known as the

Richmond Howitzers and was composed of the very flower of the young men of that city; it was their first fight, and to many their last. A member of the battery, in describing it to an officer of the sharpshooters soon after the close of the war, said pithily:

> We went in a battery and came out a wreck. We stayed ten minutes by the watch and came out with one gun, ten men and two horses, and without firing a shot.

The advanced position held by the sharpshooters being no longer tenable, as they were exposed to the fire, not only of the rebels in front but to that of their friends in the rear as well, they were withdrawn and formed in line of battle in the rear of the Fourth Michigan Volunteers, where they remained for a short time. The rebel fire from the brink of the ravine from which the sharpshooters had been dislodged, as before described, now became exceedingly galling and troublesome to the artillery in our front line, and several horses and men were hit in Weeden's R.I. battery, an officer of which requested that an effort be made to silence the fire. Col. Ripley directed Lieut. J. Smith Brown of Co. A, acting adjutant, to take twenty volunteers far out to the left and front to a point designated, which it was hoped would command the ravine. The duty was one of danger, but volunteers were quickly at hand, among whom were several from Co. F. The gallant little band soon gained the coveted position, and thereafter the fire of the rebel riflemen from that point was of little moment. Lieut. Brown's command maintained this position during the entire battle, and being squarely on the flank of Magruder's charging columns, and being, from the very smallness of their numbers, hardly noticeable among the thousands of struggling men on that fatal field, they inflicted great damage and loss in the Confederate hosts. It was now late in the afternoon, no large bodies of the rebel infantry had as yet shown themselves, though the clouds of dust arising beyond the woods told plainly of their presence and motions. A partial attack had been made on the extreme right of Morell's line, involving to some extent the left of Couch's division, but was easily repulsed; the fire of Co. E of the sharpshooters, which had been sent to that point, contributing largely to that result. The artillery fire had been heavy and incessant for some hours, and shells were bursting in quick succession over every portion of the field. Suddenly there burst out of the ravine a heavy line of battle, followed by another and another, while out of the woods beyond poured masses of men in support. The battle now commenced in earnest.

The Union infantry, heretofore concealed and sheltered behind such little inequalities of ground as the field afforded, sprang to their feet and opened a tremendous fire, additional batteries were brought up, and from every direction shot and shell, canister and grape, were hurled against the advancing enemy, while the gun boats, at anchor in the river two miles away, joined their efforts with those of their brethren of the army. It was a gallant attempt, but nothing human could stand against the storm—great gaps began to be perceptible in the lines, but the fiery energy of Magruder was behind them and they still kept on, until it seemed that nothing short of the bayonet would stop them. Gradually, however, the rush was abated; here and there could be seen signs of wavering and hesitation; this was the signal for redoubled efforts on the part of the Union troops, and the discomfited rebels broke in confusion and fled to the shelter of the woods and ravines.

At the critical moment of this charge the sharpshooters had been thrown into line on the right of the fourth Michigan regiment and bore an honourable part in the repulse; indeed, so closely crowded were the Union lines at this point that many men of the sharpshooters found themselves in the line of the Michigan regiment and fought shoulder to shoulder with their western brothers. The battle was, however, by no means over; again and again did Magruder hurl his devoted troops against the Union line, only to meet a like repulse; the rebels fought like men who realized that their efforts of the past week, measurably successful though they had been, would have failed of their full result should they now fail to destroy the Army of the Potomac; while the Union troops held their lines with the tenacity of soldiers who knew that the fate of a nation depended upon the result of that day. At the close of the second assault the sharp-shooters found themselves with empty cartridge boxes and were withdrawn from the front. The special ammunition required for their breech loaders not being obtainable, they were not again engaged during the day. In this fight the regiment lost many officers and men, among whom were Col. Ripley, Capt. Austin and Lieut. Jones of Co. E, wounded. In Co. F, Lieut. C. W. Seaton, Jacob S. Bailey and Brigham Buswell were wounded. Buswell's wound resulted in his discharge. Bailey rejoined the company, only to lose an arm at Chancellorsville.

The final rebel attack having been repulsed and their defeat being complete and final, the Union army was withdrawn during the night to Harrison's Landing, some eight miles distant, which point had been selected by Gen. McClellan's engineers some days before as the base for future operations against Richmond by the line of the James River; operations which, as the event proved, were not to be undertaken until after two years of unsuccessful fighting in other fields, the Army of the Potomac found itself once more on the familiar fields of its earliest experience. The Campaign of the Peninsula was over; that mighty army that had sailed down the beautiful Potomac so full of hope and pride less than four months before; that had through toil and suffering fought its way to within sight of its goal; found itself beaten back at the very moment of its anticipated triumph, and instead of the elation of victory, it was tasting the bitterness of defeat; for, although many of its battles, as that of Hanover Court House, Williamsburgh, Yorktown, Mechanicsville and Malvern Hill, had been tactical victories, it felt that the full measure of success had not been gained, and that its mission had not been accomplished. While the army lay at Harrison's Landing the following changes in the rolls of Co. F took place: Sergeant Amos H. Bunker, Azial N. Blanchard, Wm. Cooley, Geo. W. Manchester and Chas. G. Odell were discharged on surgeon's certificate of disability, and Brigham Buswell was discharged on account of disability resulting from the wound received at Malvern Hill. Benajah W. Jordan and James A. Read died of wounds received at Gaines Hill and W. S. Tarbell of disease. E. F. Stevens and L. D. Grover were promoted sergeants, and W. H. Leach and Edward Trask were made corporals. At this camp also Capt. Weston resigned and Lieut. C. W. Seaton was appointed captain, Second Lieut. M. V. B. Bronson was promoted first lieutenant and Ezbon W. Hindes second lieutenant. Major Trepp was promoted lieutenant-colonel, vice Wm. Y. W. Ripley, and Capt. Hastings of Co. H. was made major.

The regiment remained at Harrison's Landing until the army left the Peninsula. The weather was intensely hot and the army suffered terrible losses by disease, cooped up as they were on the low and unhealthy bottom lands bordering the James. The enemy made one or two demonstrations, and on one occasion the camp of the sharpshooters became the target for the rebel batteries posted on the high lands on the further side of the river, and for a long time the men of Co. F were exposed to a severe fire to which they could not reply, but luckily without serious loss.

CHAPTER 3

Second Bull Run, Antietam, Fredericksburgh

About the middle of August, the government having determined upon the evacuation of the Peninsula, the army abandoned its position at Harrison's Landing. Water transportation not being at hand in sufficient quantity, a large portion of the army marched southward towards Fortress Monroe, passing, by the way, the fields of Williamsburgh. Lee's Mills and Yorktown, upon which they had so recently stood victorious over the very enemy upon whom they were now turning their backs. Co. F. was with the division which thus passed down by land. Upon arriving at Hampton the Fifth Corps, to which the sharpshooters were attached, embarked on steamboats and were quickly and comfortably conveyed to Acquia Creek, at which place they took the cars for Falmouth, on the Rappahannock opposite Fredericksburgh.

No sooner did McClellan turn his back on Richmond in the execution of this change of base, than Lee, no longer held to the defence of the rebel capital, moved with his entire force rapidly northward, hoping to crush Pope's scattered columns in detail before the Army of the Potomac could appear to its support. Indeed, before McClellan's movement commenced, the Confederate General Jackson—he whose foray in the valley in May had so completely neutralized McDowell's powerful corps that its services were practically lost to the Union commander during the entire period of the Peninsular Campaign—had again appeared on Pope's right and rear, and it was this apparition that struck such dread to the soul of Halleck, then General-in-Chief at Washington. Now commenced that campaign of manoeuvres in which Pope was so signally foiled by his keen and wary antagonist.

The Fifth Corps left Falmouth on the 24th of August, marching to Rappahannock Station, thence along the line of the Orange & Alexandria railroad to Warrenton Junction where they remained for a few hours, it being the longest rest they had had since leaving Falmouth, sixty miles away. On the 28th of August the sharpshooters arrived, with the rest of the corps, at Bristoe's Station where Porter had been ordered to take position at daylight to assist in the entertainment which Pope had advertised for that day, and which was to consist of "*bagging the whole crowd*" of rebels.

The wily Jackson, however, was no party to that plan, and while Pope was vainly seeking him about Manassas Junction, he was quietly awaiting the arrival of Lee's main columns near Groveton. The corps remained at Bristoe's, or between that place and Manassas Junction, inactive during the rest of the twenty-eighth and the whole of the twenty-ninth, and the sharpshooters thus failed of any considerable share in the battle of Groveton on that day. During the night preceding the 30th of August, Porter's corps was moved by the Sudley Springs road from their position near Bristoe's to the scene of the previous day's battle to the north and east of Groveton, where its line of battle was formed in a direction nearly northeast and southwest, with the left on the Warrenton turnpike. Morell's division, to which the sharpshooters were attached, formed the front line with the sharpshooters, as usual, far in the advance as skirmishers. With a grand rush the riflemen drove the rebels through the outlying woods, and following close upon the heels of the flying enemy, suddenly passed from the comparative shelter of the woods into an open field directly in the face of Jackson's corps strongly posted behind the embankment of an unfinished railroad leading from Sudley Springs southwestwardly towards Groveton.

It was a grand fortification ready formed for the enemy's occupation, and stoutly defended by the Stonewall brigade. Straight up to the embankment pushed the gallant sharpshooters, and handsomely were they supported by the splendid troops of Barnes and Butterfield's brigades. The attack was made with the utmost impetuosity and tenaciously sustained; but Jackson's veterans could not be dislodged from their strong position behind their works. The sharpshooters gained the shelter of a partially sunken road parallel to the enemy's line and hardly thirty yards distant; but not even the splendid courage of the men who had held the lines of Gaines Hill and Malvern against this same enemy, could avail to drive them from their shelter.

To add to the peril of the charging column, Longstreet, on Jackson's right, organized an attack on Porter's exposed left flank. The corps thus placed, with an enemy in their front whom they could not dislodge and another on their unprotected flank, were forced to abandon their attack. The sharpshooters were the last to leave their advanced positions, and then only when, nearly out of ammunition, Longstreet's fresh troops fairly crowded them out by sheer numerical superiority. Of Co. F the following men were wounded in this battle: Corporals H. J. Peck and Ai Brown and Private W. H. Blake. Corporal Peck was honourably discharged on the 26th of October following for disability resulting from his wound. The sharpshooters were not again seriously engaged with the enemy during Pope's campaign. On the night after the battle they retired with the shattered remains of the gallant Fifth Corps, and on the 1st of September went into camp near Fort Corcoran. So far the campaigns of the sharpshooters had, although full of thrilling incident and gallant achievement, been barren of result. Great victories had been won on many fields, but the end seemed as far off as when they left Washington more than five months before.

Disease and losses in battle had sadly thinned their ranks, but the remnant were soldiers tried and tempered in the fire of many battles. They were not of the stuff that wilts and shrivels under an adverse fortune, and putting the past resolutely behind them, they set their faces sternly towards the future, prepared for whatever of good, or of ill, it should have in store for them.

THE ANTIETAM CAMPAIGN

On the 12th of September, the main portion of the army having preceded them, the Fifth Corps crossed to the north bank of the Potomac, and by forced marches came up with the more advanced columns on the sixteenth and took part in the manoeuvres which brought the contending armies again face to face on the banks of the Antietam.

The rebels, flushed with the very substantial advantages they had gained during the past summer, were confident and full of enthusiasm. Posted in an exceptionally strong position, their flanks resting on the Potomac while their front was covered by the deep and rapid Antietam, they calmly awaited the Union attack, confident that the army which they had so signally discomfited under Pope would again recoil before their fire. But the Union situation was not the same that it had been a month before; McClellan had resumed the command, not only of the old Army of the Potomac—the darling child of his own creation, and which in turn loved and honoured him with a devotion difficult for the carping critic of these modern times to understand—but of the remains of the army of Northern Virginia as well.

These incongruous elements he had welded together, reorganized and re-equipped while still on the march, until, when they stood again before Lee's hosts on the banks of Antietam creek on the 17th of September, they were as compact in organization and as confident as at any previous time in their history. Then, too, they were to fight on soil which, if not entirely loyal, was at least not the soil of the so called Confederate States; and the feeling that they were called upon for a great effort in behalf of an endangered North, gave an additional stimulus to their spirits and nerved their arms with greater power. But with the history of this great battle we have little to do. The Fifth Corps was held in reserve during the entire day. It was the first time in the history of the company that its members had been lookers on while rebel and Unionist fought together; here, however, they could, from their position, overlook most of the actual field of battle as mere spectators of a scene, the like of which they had so often been actors in.

On the day after the battle they received a welcome addition to their terribly reduced ranks by the arrival of some fifty recruits under Lieut. Bronson, who had been detached on recruiting service while the army yet lay along the Chicahominy during the previous month of June. On the 19th of September the pursuit of Lee's retreating army was taken up, the Fifth Corps in the advance, and the sharpshooters leading the column. The rear guard of the enemy was overtaken at Blackford's Ford, at which place Lee had re-crossed the Potomac.

The rebel skirmishers having been driven across the river, preparations for forcing the pursuit into Virginia were made, and the sharpshooters were ordered to cross and drive the rebel riflemen from their sheltered positions along the Virginia shore. The water was waist deep but, holding their cartridge boxes above their heads, they advanced in skirmish line totally unable to reply to the galling fire that met them as they entered the stream. Stumbling and floundering along, they at last gained the farther shore and quickly succeeded in compelling the rebels to retire.

Advancing southward to a suitable position, Co. F was ordered to establish an advanced picket line in the execution of which order a party under Corporal Cassius Peck discovered the presence of a small body of the enemy with two guns, who had been left behind for some reason by the retreating rebels. This force was soon put to flight and both guns captured and one man taken prisoner. The captured guns were removed to a point near the river bank, from which they were subsequently removed to the Maryland shore. Remaining in this position until after dark the sharpshooters were ordered back to the north bank of the river, to which they retired. Morning found them posted in the bed of the canal which connects Washington with Harper's Ferry, and which runs close along the Maryland shore of the Potomac at this point. The water being out of the canal, its bed afforded capital shelter, and its banks a fine position from which to fire upon the rebels, now again in full possession of the opposite shore from which they had been driven by the sharpshooters the previous afternoon, but which had been deliberately abandoned to them again by the recall of the regiment to the northern shore on the preceding night.

It now became necessary to repossess that position, and a Pennsylvania regiment composed of new troops were ordered to make the attempt. Covered by the close and rapid fire of the sharpshooters, the Pennsylvanians succeeded in crossing the river, but every attempt to advance from the bank met with repulse. Wearied and demoralized by repeated failures, the regiment took shelter under the banks of the river where they were measurably protected from the fire of the enemy, and covered also by the rifles of the sharpshooters posted in the canal. Ordered to re-cross the river, they could not be induced by their officers to expose themselves in the open stream to the fire of the exulting rebels.

Every effort was made by the sharpshooters to encourage them to re-cross, but without avail. Calvin Morse, a bugler of Co. F, and thus a non-combatant (except that Co. F had no non-combatants), crossed the stream, covered by the fire of his comrades, to demonstrate to the panic stricken men that it could be done; but they could not be persuaded, and most of them were finally made prisoners. In these operations Co. F was exceptionally fortunate, and had no casualties to report.

The regiment remained at or near Sharpsburgh, Maryland, until the 30th of October following. The members of Co. F, except the recruits, were but poorly supplied with clothing; much had been abandoned and destroyed when they left their camp at Gaines Hill on the 27th of June, and much, also, had been thrown away to lighten the loads of the tired owners during the terrible marches and battles they had passed through since that time, and the little they had left was so worn and tattered as to be fit for little more than to conceal their nakedness. The rations, too, were bad; the hard bread particularly so, being wormy and mouldy, and this at a place and time when it seemed to the soldiers that there could be no good reason why such a state of things should exist at all. But time cures all ills, even in the army, and on the 30th of October the regiment, completely refitted, rested and in fine spirits, crossed the Potomac at Harper's Ferry and were once more on the sacred soil of Virginia. Moving southwardly towards Warrenton they arrived, on the evening of November 2nd, at Snicker's Gap and were at once pushed out to occupy the summit. The night was intensely dark, and the ground difficult; but a proper picket line was finally established and occupied without event through the night. The next morning's sunlight displayed a wonderful sight to the eyes of the delighted sharpshooters. They were on the very summit of the Blue Ridge Mountains, and below them, like an open map, lay spread out the beautiful valley of Virginia.

Scathed and torn as it was, to a close observer, by the conflicts and marches of the past summer, from the distant point of view occupied by the watchers, all was beautiful and serene. No sign of war, or its desolating touch, was visible; except that here and there could be seen bodies of marching men, and long trains of wagons, which told of the presence of the enemy. Now, however, the head of every column was turned southward, and the rebel army, which had swept so triumphantly northward over that very country only two months before, was retiring, beaten and baffled, before the army of the Union. The scene was beautiful to the eye, while the reflections engendered by it were of the most hopeful nature, and the sharpshooters descended the southern slope of the mountain with high hopes and glowing anticipations of speedy and decisive action.

From Snicker's Gap the army advanced by easy marches to Warrenton, where, on the 7th of November, Gen. McClellan was relieved from the command and Gen. Burnside appointed to that position. The army accepted the change like soldiers, but with a deep sense of regret. The vast mass of the rank and file honoured and trusted Gen. McClellan as few generals in history have been trusted by their followers. He was personally popular among the men, but below and behind this feeling was the belief that in many respects Gen. McClellan had not been quite fairly treated by some of those who ought to have been his warm and ardent supporters. They felt that political influences, which had but little hold upon the soldiers in the field, had been at work to the personal disadvantage of their loved commander, and to the disadvantage of the army and the cause of the Union as well.

Whether they were right or wrong, they regretted the change most deeply, and in this general feeling the sharpshooters stood with the great mass of the army.

While they were always ready with a prompt obedience and hearty support of their later commanders, the regiment never cheered a general officer after McClellan left the head of the Army of the Potomac.

After a few days of rest at Warrenton to allow Gen. Burnside to get the reins well in hand, the army was put in motion towards Fredericksburgh where they arrived on or about the 23rd of November. While at Warrenton Gen. Burnside effected a complete reorganization of the army, on a plan which he had been pressing upon the notice of his superiors for some time. The entire army was divided into three Grand Divisions, the right under Sumner, the centre under Hooker, and the left under Franklin. The Fifth Corps formed part of the Centre Grand Division under Gen. Hooker, and at about the same time Gen. F. J. Porter, who had been its commander since its organization while the army lay before Yorktown during the preceding April, was relieved from his command and was succeeded by Gen. Dan'l Butterfield.

Gen. Burnside, having been disappointed in finding his pontoon trains, on which he depended for a rapid passage to the south bank of the Rappahannock, ready on his arrival at Falmouth, was constrained to attempt to force a passage in the face of Lee's now concentrated army. The position was one well calculated to dampen the ardour of the troops now so accustomed to warfare as to be able to weigh the chances of success or failure as accurately as their commanders, and to judge quickly of the value to their cause of that for which they were asked to offer up their lives, but they undertook the task as cheerfully and as willingly as though it had been far less uncertain and perilous. The Rappahannock at this point is bordered by opposing ranges of hills; that on the left bank, occupied by the troops of the Union and called Stafford Heights, rising quite abruptly from the river bank; while on the southern shore the line of hills, called Marye's Heights, recedes from the river from six hundred to two thousand yards, the intervening ground being generally open and, although somewhat broken, affording very little shelter from the fire of the Confederate batteries posted on Marye's Heights. On the plain and near the river stands the village of Fredericksburgh.

During the night of the 10th of December Gen. Burnside placed in position on Stafford Heights a powerful array of guns, under cover of whose fire he determined to attempt the passage of the river at that point, while to the Left Grand Division under Franklin was assigned the task of forcing a passage at a point some two miles lower down. On the night of the 11th attempts were made to lay the pontoon bridges at a point opposite the town. The enemy, however, well warned, posted a strong force of riflemen in the houses and behind the stone walls bordering the river, whose sharp fire so seriously impeded the efforts of the engineers that they were forced to retire. The guns on Stafford Heights were opened on the town, and for nearly two hours one hundred and fifty guns poured their shot and shell upon the devoted town. Each gun was estimated to have fired fifty rounds; but at the close of the bombardment the annoying riflemen were still there. Three regiments were now thrown across the river in pontoon boats, and after a severe fight in the streets of the town, and after heavy loss of men, succeeded in dislodging the enemy, and the bridges were completed. Of course a surprise, upon which Burnside seems to have counted, was now out of the question; but urged on by the voice of the North, whose sole idea at that time seemed to be that their generals should only fight—anywhere, under all circumstances and at all times—he threw Sumner's Grand Division over the river and determined to try the issue of a general battle.

The Centre Grand Division, under Hooker, were held on the left bank of the river and were thus unengaged in the earlier portion of that terrible day; but from their position on Stafford Heights, the sharpshooters were eye witnesses to the terrible struggle in which their comrades were engaged on the plain below—where Hancock's gallant division, in their desperate charge upon the stone wall at the foot of Marye's Heights, lost two thousand men out of the five thousand engaged in less than fifteen immortal minutes, and where a total of twelve thousand, three hundred and twenty-nine Union soldiers fell in the different assaults; assaults that every man engaged knew were utterly hopeless and vain; but to the everlasting honour of the Army of the Potomac be it said that, although they well knew the task an impossible one, they responded again and again to the call to advance, until Burnside himself, at last convinced of the hopelessness of the undertaking, suspended further effort.

During the day Griffin and Humphrey's divisions of the Fifth Corps, and Whipple's of the Third, all belonging to the Centre Grand Division, were ordered over the river to renew the attack which had been so disastrous to the men of the Second and Ninth Corps. Hooker in person accompanied this relieving column, and after a careful personal inspection of the field, convinced of the uselessness of further effort in that direction, sought to persuade the commanding general to abandon the attack.

Burnside, however, clung to the hope that repeated attacks must at last result in a disruption of the enemy's line at some point, and the brave men of the old Fifth were in their turn hurled against that position which had been found impossible to carry by those who had preceded them. Griffin and Humphrey's divisions fought their way to a point farther advanced than had been reached in former attempts, some of the men falling within twenty-five yards of the enemy's line, but they were unable to reach it and were compelled to retire. It was clearly impossible to carry the position. Hooker's educated eye had seen this from the first, hence his unavailing suggestion before the useless slaughter. His report contains the following grim lines:

Finding that I had lost as many men as my orders required me to lose, I suspended the attack.

With his repulse the battle of Fredericksburgh substantially closed. The sharpshooters were not ordered to cross the river on the thirteenth, and thus had no share in that day's fighting and no casualties to report. On the early morning of the fourteenth, however, the remainder of the Centre Grand Division crossed to the south bank, remaining in the streets of the town until the night of the fifteenth, when the sharpshooters relieved the advanced pickets in front of the heights, where considerable firing occurred during the night, the opposing lines being very near each other. The ground was thickly covered with the bodies of the gallant men who had fallen in the several assaults, lying in every conceivable position on the field, gory and distorted. How many of the readers of this book will make it real to themselves what gore is? A familiar and easily spoken word, but a dreadful thing in reality, that mass of clotted, gelatinous purple oozing from mortal wounds.

Such things are rarely noted in the actual heat of the battle, but to occupy such a field after the fury of the strife is over is enough to unman the stoutest heart, and many a brave man, who can coolly face the actual danger, turns deathly sick as he looks upon the result as shown in the mangled and blood stained forms of those who were so lately his comrades and friends. During the night the army was withdrawn to the north bank, and just before daylight the sharpshooters were called in. So close were the lines that great caution was necessary to keep the movement from the sharp eyes of the peering rebel pickets. To aid in deceiving the enemy the bodies of the dead were propped up so as to represent the presence of the picket line when daylight should appear. The ruse was successful, and the sharpshooters were safely withdrawn to the town. They were the last troops on this portion of the field, and on arriving at the head of the bridge found that the planking had been so far removed as to render the bridge impassable. They had, therefore, to remain until the engineers could relay sufficient of the planks to enable them to cross. In their retreat through the town they picked up and brought away about one hundred and fifty stragglers and slightly wounded men who had been left behind by other commands. The Army of the Potomac was again on the north bank of the Rappahannock. They had fought bravely in an assault which they had known was hopeless; they had left behind them twelve thousand of their comrades and gained absolutely nothing. The loss which they had inflicted bore no proportion to that which they had suffered; what wonder, then, if for a time officers and men alike almost despaired of the cause of the Union? This feeling of depression and discouragement was, however, of short duration. The men who composed the Army of the Potomac were in the field for a certain well defined purpose, and until that purpose was fully accomplished they intended to remain. No reverse could long chill their ardour or dampen their splendid courage. Defeated today, tomorrow would find them as ready to do and dare again as though no reverse had overtaken them.

Thus it was that after a few days of rest the army was ready for whatever task its commander might set for it. The sharpshooters remained quietly in their camp until the 30th of December, when they accompanied a detachment of cavalry on a reconnaissance northwardly along the line of the Rappahannock to Richard's Ford, some ten miles above Falmouth. The cavalry crossed the river at this point, covered by the fire of the sharpshooters; a few prisoners were taken, and on the 1st of January, 1863, the command returned to their comfortable camp near Falmouth, where they were agreeably surprised to find the Second Regiment of Sharpshooters, and among them, two other companies from Vermont. The little band of Green Mountain boys composing Co. F had sometimes felt a little lonesome for the want of congenial society, and hailed the advent of their fellow Vermonters gladly.

At about this time Col. Berdan became an appendage to the general staff, with the title of chief of sharpshooters. The two regiments were distributed at various points along the line, and the detachments reported directly to Col. Berdan. The right wing, under Lieut. Col. Trepp, was assigned to the Right Grand Division under Gen. Sumner, but Company F remained near army headquarters.

On the 19th of January the Grand Divisions of Franklin and Hooker moved up the river to essay its passage at Banks' Ford, some six miles above Falmouth, but in this affair, known as the Mud Campaign, the company had no share, not even leaving their camp. Of this campaign it is enough to say that it had for its object a turning operation similar to that undertaken by Hooker some months later; but a furious rain storm converted the country into one vast quagmire, in which horses, wagons, guns and men were alike unable to move. It was entirely abortive, and, after two days of exhausting labour, the disgusted troops floundered and staggered and cursed their way back to their camps, actually having to build corduroy roads on which to return. In consideration of their dry and comfortable condition in camp, the sharpshooters freely conceded all the glories of this campaign to others, preferring for themselves an inglorious ease to the chance of being smothered in the mud. Some of the difficulties of the march can be understood by recalling the requisition of the young engineer officer who reported to his superior that it was impossible for him to construct a road at a certain point which he had been directed to make passable for artillery. "Impossible," said the commander, "nothing is impossible; make a requisition for whatever is necessary and build the road." Whereupon the officer made the following requisition in the usual form:

Special Requisition

Requisition for Men

Fifty men, each twenty-five feet high, to work in the mud eighteen feet deep.

I certify that the above described men are necessary to the building of a road suitable for the passage of men and guns, in compliance with an order this day received from Major-Gen.——.

Signed,

————,

Lieut. Engineers

On the 25th of January Gen. Burnside was relieved from the command and Gen. Hooker appointed to succeed him. The army accepted the change willingly, for although they recognized the many manly and soldierly qualities possessed by Gen. Burnside, and in a certain way respected and even sympathized with him, they had lost confidence in his ability to command so large an army in the presence of so astute a commander as Lee. His manly avowal of his sole responsibility for the terrible slaughter at Fredericksburgh commended him to their hearts and understandings as an honest and generous man; but they had no wish to repeat the experience for the sake of even a more generous acknowledgement after another Fredericksburgh.

The remainder of the winter of 1862-3 was spent by the men of Co. F in comparative comfort, although severe snow storms were of frequent occurrence, and occasional periods of exceedingly cold weather were experienced, to the great discomfort of the men in their frail canvas tents. Both armies seemed to have had enough of marching and fighting to satisfy them for the time being, and even picket firing ceased by tacit agreement and consent.

Soon after assuming command, Gen. Hooker reorganized the army on a plan more consistent with his own ideas than the one adopted by his predecessor. The system of Grand Divisions was abandoned and corps were reorganized; some corps commanders were relieved and others appointed to fill the vacancies. The cavalry, which up to this time had had no organization as a corps, was consolidated under Gen. Stoneman, and soon became, under his able leadership, the equals, if not the superiors, of the vaunted horsemen of the South. In these changes the sharpshooters found themselves assigned to the first division of the Third Corps, under Gen. Sickles. The division was commanded by Gen. Whipple, and the brigade by Gen. De Trobriand. The detachments were called in and the regiment was once more a unit. Under Gen. Hooker's system the army rapidly improved in morale and spirit; he instituted a liberal system of furloughs to deserving men, and took vigorous measures against stragglers and men absent without leave, of whom there were at this time an immense number—shown by the official rolls to be above eighty thousand. Desertion, which under Burnside had become alarmingly prevalent, was substantially stopped; and by the 1st of April the tone and discipline of the army was such as to fairly warrant Hooker's proud boast that it was "the grandest army on the planet."

The sharpshooters parted with their comrades of the Fifth Corps with regret. They had been identified with it since its organization, while the army lay before Yorktown, in April of 1862; they had shared with it splendid triumphs and bitter defeats; they had made many warm friends among its officers and men, with whom they were loath to part. Of the officers of the Third Corps they knew nothing, but they took their place in its ranks, confident that their stout soldiership would win for them the respect and esteem of their new comrades, even as it had that of the friends they were leaving. Gen. De Trobriand, their new brigade commander, was at first an object of special aversion. Foreign officers were at that time looked upon with some degree of suspicion and dislike, and perhaps the foreign sound of the name, together with the obnoxious prefix, had an undue and improper influence in the minds of the new comers. However it came about, the men were accustomed to speak of their superior officer as Gen. "Toejam," "Frog Eater," and various other disrespectful appellations, much to his chagrin and discomfiture. Later, however, when they became better acquainted, they learned to have a mutual respect and esteem for each other and two years later, when they parted company finally, the general issued to them a farewell address more than usually complimentary, as will be seen further on. Indeed, long before that time and on the field of actual and bloody battle he paused in front of the line of the regiment to say to them: "Men, you may call me *Frog Eater* now if you like, or by whatever name you like better, if you will only always fight as you do today."

The sharpshooters passed the winter months in comparative inaction except for the ordinary routine of drills, inspections, etc., incident to winter quarters; they took part in all the grand reviews and parades for which Hooker was somewhat famous, and which, if somewhat fatiguing to the men and smacking somewhat of pomp and circumstance, had at least the effect of showing to each portion of the great army what a magnificent body they really were, thus adding to the confidence of the whole.

On the twenty-first of February First Lieut. Bronson resigned, and was succeeded by Lieut. E. W. Hindes, while, in deference to the unanimous petition of the company, Sergt. C. D. Merriman was promoted second lieutenant, both commissions to date from February 21, 1863. The roster of the company now stood as follows:

Captain	C. W. Seaton.
First Lieutenant	E. W. Hindes.
Second Lieutenant	C. D. Merriman.
First Sergeant	H. E. Kinsman.
Second Sergeant	A. H. Cooper.
Third Sergeant	Cassius Peck.
Fourth Sergeant	Edward F. Stevens.
Fifth Sergeant	Lewis J. Allen.
First Corporal	Paul M. Thompson.
Second Corporal	Ai Brown.
Third Corporal	L. D. Grover.
Fourth Corporal	Chas. M. Jordan.

Fifth Corporal	E. M. Hosmer.
Sixth Corporal	Edward Trask.
Seventh Corporal	W. H. Leach.
Eighth Corporal	M. Cunningham.

The winter was not altogether devoted to sober work. Sports of various kinds were indulged in, one of the most popular being snowball fights between regiments and brigades. Upon one occasion after a sharp conflict between the first and second regiments of sharpshooters, the former captured the regimental colours of the latter, and for a short time some little ill feeling between the regiments existed, a feeling which soon wore away, however, with the opening of the spring campaign.

On the 5th of April the first regiment had a grand celebration to mark the anniversary of the advance on Yorktown where the sharpshooters were for the first time under rebel fire. Target shooting, foot races, jumping and wrestling were indulged in for small prizes. Jacob S. Bailey of Co. F won the wrestling match against all comers and Edward Bartomey, also of Company F, won the two hundred yards running race in twenty-eight and one-half seconds. In the shooting test the Vermonters were unfortunate, the prize going to Samuel Ingling of Michigan. Gen. Whipple, the division commander, accompanied by several ladies who were visiting friends in camp, were interested spectators of the games. As the season advanced and the roads became settled and passable, preparations began on all sides for an active campaign against the enemy. "Fighting Joe Hooker" had inspired the army with much of his own confidence and faith in the future, and it was believed by the troops that at last they had a commander worthy in every respect of the magnificent army he was called to command.

Chapter 4

Chancellorsville

On the 28th of April the Third Corps, to which the sharpshooters were now attached, moved down the river to a point some five miles below Falmouth to support Sedgwick's command which was ordered to cross the Rappahannock at or near the point at which Gen. Franklin had crossed his Grand Division at the battle of Fredericksburgh.

Some days prior to this all surplus clothing and baggage had been turned in. Eight days rations and sixty rounds of ammunition were now issued, and the *"finest army on the planet"* was foot loose once more. Sedgwick's crossing was made, however, without serious opposition, and on the thirtieth the Third Corps, making a wide detour to the rear to avoid the notice of the watchful enemy, turned northward and on the next day crossed the river at United States Ford and took its place in the lines of Chancellorsville with the rest of the army.

This great battle has been so often described and in such minute detail that it is not necessary for us to attempt a detailed description of the movements of the different corps engaged, or indeed proper, since this purports to be a history of the marches and battles of only one small company out of the thousands there engaged. It will be remembered that the regiment was now attached to the Third Corps, commanded by Gen. Sickles, the First Division under Gen. Whipple and the Third Brigade, Gen. De Trobriand. At eleven o'clock a.m. on this day, being the first of May, the battle proper commenced, although severe and continuous skirmishing had been going on ever since the first troops crossed the river on the 29th of April. The Third Corps was held in reserve in rear of the Chancellorsville house, having arrived at that point at about the time that the assaulting columns moved forward to the attack. Almost instantly the fighting became furious and deadly. The country was covered with dense undergrowth of stunted cedars, among and over which grew heavy masses of the trailing vines which grow so luxuriantly in that portion of Virginia, and which renders the orderly passage of troops well nigh impossible.

To add to the difficulties which beset the attacking forces, it was impossible to see what was in front of them; hence the first notice of the presence of a rebel line of battle was a volley delivered at short range directly in the faces of the Union soldiers, whose presence and movements were unavoidably made plain to the concealed enemy by the noise made in forcing a passage through the tangled forest. Notwithstanding these disadvantages the Fifth Corps, with which the sharpshooters had so recently parted, struck the enemy at about a mile distant from the position now held by the Third Corps, and drove them steadily back for a long distance until, having passed far to the front of the general line, Meade found his flank suddenly attacked and was forced to retire. Other columns also met the enemy at about the same distance to the front and met with a like experience, gaining, however, on the whole, substantial ground during the afternoon; and so night closed down on the first day of the battle.

On the morning of the 2nd of May a division of the Third Corps was detached to hold a gap in the lines between the Eleventh and Twelfth Corps which Gen. Hooker thought too weak. The sharpshooters, however, remained with the main column near the Chancellorsville house. Early on this day the Confederate Gen. Jackson commenced that wonderful flank march which resulted in the disaster to the Eleventh Corps on the right, later in the day. This march, carefully masked as it was, was, nevertheless, observed by Hooker, who at first supposed it the commencement of a retreat on the part of Lee to Gordonsville, and Gen. Sickles was ordered with the two remaining divisions of his corps to demonstrate in that direction and act as circumstances should determine. In this movement Birney's division had the advance, the first division, under Whipple, being in support of Birney's left flank. The sharpshooters were, however, ordered to report to Gen. Birney, and were by him placed in the front line as skirmishers, although their deployment was at such short intervals that it was more like a single rank line of battle than a line of skirmishers. Sickles started on his advance at about one o'clock p.m., his formation being as above described. Rapidly pressing forward, the sharpshooters passed out of the dense thickets into a comparatively open country, where they could at least breathe more freely and see a little of what was before them. They soon struck a line of rebels in position on the crest of a slight elevation, and brisk firing commenced; the advance, however, not being checked, they soon cleared the hill of the enemy and occupied it themselves. Changing front to the left, the regiment moved from this position obliquely to the southeast, and soon found themselves opposed to a line which had evidently come to stay. The fighting here was very severe and lasted for a considerable time. The rebels seemed to have a desire to stay the advance of the Union troops at that particular point, and for some particular reason, which was afterwards made apparent.

After some minutes of brisk firing, the sharpshooters, by a sudden rush on their flank, succeeded in compelling the surrender of the entire force, which was found to consist of the Twenty-third Georgia regiment, consisting of three hundred and sixty officers and men, which had been charged by Jackson with the duty of preventing any advance of the Union troops at this point which might discover his march towards Hooker's right, hence the tenacity with which they clung to the position.

In this affair Co. F lost Edward Trask and A. D. Griffin, wounded.

The obstruction having been thus removed, the Third Corps, led by the sharpshooters, pressed rapidly forward to the southward as far as Hazel Grove, or the old furnace, some two miles from the place of starting, and far beyond any supporting column which could be depended on for early assistance should such be needed. It had now become apparent to all that Jackson, instead of being in full retreat as had been supposed, was in the full tide of one of the most violent offensives on record; and at five o'clock p.m. Sickles was ordered to attack his right flank and thus check his advance on the exposed right of the army. But at about the same time Sickles found that he was himself substantially cut off from the army, and that it would require the most strenuous efforts to prevent the capture or destruction of his own command. Furthermore, before he could make his dispositions and march over the ground necessary to be traversed before he could reach Jackson's right, that officer had struck his objective point, and the rout of the Eleventh Corps was complete. The most that Sickles could now do, under the circumstances, was to fight his own way back to his supports, and to choose, if possible, such a route as would place him, on his arrival, in a position to check Jackson's further advance and afford the broken right wing an opportunity to rally and regain their organization, which was hopelessly, as it appeared, lost. In the darkness and gloom of the falling night, with unloaded muskets (for in this desperate attempt the bayonet only was to be depended upon), the two divisions of the Third Corps set their faces northwardly, and pressed their way through the tangled undergrowth to the rescue of the endangered right wing.

As usual, the sharpshooters had the advance, and received the first volley from the concealed enemy. They had received no especial orders concerning the use, solely, of the bayonet, and were at once engaged in a close conflict under circumstances in which their only superiority over troops of the line consisted in the advantage of the rapidity of fire afforded by their breech loaders over the muzzle loading rifles opposed to them. Closely supported by the line of Birney's division, and firing as they advanced at the flashes of the opposing guns (for they could see no more), they pushed on until they were fairly intermingled with the rebels, and in many individual instances, a long distance inside the enemy's line, every man fighting for himself—for in this confused melee, in the dense jungle and in the intense darkness of the night, no supervision could be exercised by officers and many shots were fired at distances no greater than a few feet. So they struggled on until, with a hurrah and a grand rush, Birney's gallant men dashed forward with the bayonet alone, and after ten minutes of hand to hand fighting, they succeeded in retaking the plank road, and a considerable portion of the line held by the left of the Eleventh Corps in the early portion of the day and lost in the tremendous charge of Jackson's corps in the early evening. Sickles had cut his way out, and more, he was now in a position to afford the much needed aid to those who so sorely required it. Both parties had fought to the point of exhaustion, and were glad to suspend operations for a time for this cause alone, even had no better reasons offered. But the Union army was no longer in a position for offense; the extreme left, with which we have had nothing to do, had been so heavily pressed during the afternoon that it had been with difficulty that a disaster similar to the one which had overtaken the right had been prevented on that flank, and in the centre, at and about Hazel Grove and the furnace, which had been held by Sickles, and from which he had been ordered to the support of the right as we have seen, an absolute gap existed, covered by no force whatever. This, then, was the situation, briefly stated.

The left was barely able to hold its own, the centre was absolutely abandoned, and the right had been utterly routed. In this state of affairs the Union commander was in no mood for a further offense at that time. On the other hand, the controlling mind that had conceived, and thus far had successfully carried out this wonderful attack which had been so disastrous to the Union army, and which bade fair to make the Southern Confederacy a fact among the nations, had been stricken down in the full tide of its success. Stonewall Jackson had been wounded at about nine o'clock by the fire of his own men. He had passed beyond the lines of his pickets to reconnoitre the Union position, and on his return with his staff they were mistaken by his soldiers for a body of federal cavalry and he received three wounds from the effects of which he died about a week later. So fell a man who was perhaps as fine a type of stout American soldiership as any produced on either side during the war.

The sharpshooters, with the remnant of the Third Corps, passed the remainder of the night on the plank road near Dowdall's tavern. Co. F had left their knapsacks and blankets under guard near the Chancellorsville house when they advanced from that point in the morning, as had the rest of the regiment. Under these circumstances little sleep or rest could be expected even had the enemy been in less close proximity. But with the rebel pickets hardly thirty yards distant, and firing at every thing they saw or heard, sleep was out of the question. So passed the weary night of the disastrous 2nd of May at Chancellorsville.

During the night Gen. Hooker, no longer on the offensive, had been busily engaged in laying out and fortifying a new line on which he might hope more successfully to resist the attack which all knew must come at an early hour on the morning of the third. On the extreme left the troops were withdrawn from their advanced positions to a more compact and shorter line in front of, and to the south and east of the Chancellorsville house. The centre, which at sunset was unoccupied by any considerable body of Union troops, was made secure; and at daylight Sickles, with the Third Corps, was ordered to withdraw to a position indicated immediately in front of Fairview, a commanding height of land now strongly occupied by the Union artillery. It was not possible, however, to withdraw so large a body of troops from their advanced position, in the face of so watchful an enemy, without interruption. In fact, even before the movement had commenced, the enemy took the initiative and commenced the battle of that day by a furious attack upon the heights of Hazel Grove, the position so handsomely won by the Third Corps on the previous day and from which they were ordered to the relief of the Eleventh Corps at five o'clock on the preceding afternoon, as we have seen. This height of land commanded almost every portion of the field occupied by the Union army, and from it Sickles' line, as it stood at daybreak, could be completely enfiladed. This position was held by an inadequate force for its defence; indeed, as it was far in advance of the new line of battle it may be supposed that observation, rather than defence, was the duty of its occupants. They made a gallant fight, however, but were soon compelled to retire with the loss of four guns. The rebel commander, quick to see the great importance of the position, crowned the hill with thirty guns which, with the four taken from the Unionists, poured a heavy fire on all parts of the line, devoting particular attention to Sickles' exposed left and rear.

At almost the same period of time the rebels in Sickles' front made a savage attack on his line. The men of the Third Corps fought, as they always fought, stubbornly and well, but, with a force more than equal to their own in point of numbers, flushed with their success of the previous afternoon and burning to avenge the fall of Jackson, in their front, and this enormous concentration of artillery hammering away on their defenceless left, they were at last forced back to the new line in front of Fairview.

In preparation for the withdrawal contemplated, and before the rebel attack developed itself, the sharpshooters had been deployed to the front and formed a skirmish line to the north of the plank road with their left on that highway, and thus received the first of the rebel attack. They succeeded in repulsing the advance of the first line and for half an hour held their ground against repeated attempts of the rebel skirmishers to dislodge them. The position they held was one of the utmost importance since it commanded the plank road which must be the main line of the rebel approach to Fairview, the key to the new Union line, and aware of this the men fought on with a courage and determination seldom witnessed even in the ranks of that gallant regiment. After half an hour of this perilous work, the regiment on their right having given way, the sharpshooters were ordered to move by the right flank to cover the interval thus exposed, their own place being taken by still another body of infantry. Steadily and coolly the men faced to the right at the sound of the bugle, and commenced their march, still firing as they advanced. Necessarily, however, the men had to expose themselves greatly in this movement, and as necessarily their own fire was less effective than when delivered coolly from the shelter of some friendly tree, log or bank which skirmishers are so prone to seek and so loath to leave. Still the march was made in good order and in good time, for the sharpshooters had only just time to fill the gap when the rebels came on for a final trial for the mastery. For a long time the green coated riflemen clung to their ground and gave, certainly as good, as they received. But the end of the long struggle was at hand; the regiment which had taken the position just vacated by the sharpshooters was driven in confusion, and to cap the climax of misfortune, the Union artillery, observing the withdrawal of other troops, and supposing that all had been retired, opened a furious fire of canister into the woods. The sharpshooters were now in a sad case—before them a furious crowd of angry enemies, on the left the rebel artillery at Hazel Grove sweeping their lines from left to right at every discharge, while, worst of all, from the rear came the equally dangerous fire of their own friends. To retreat was as bad as to advance. The ground to their right was an unknown mystery and no hopeful sign came from the left; so taking counsel from their very desperation they concluded to remain just there, at least until some reasonable prospect of escape should present itself. Taking such cover as they could get, some from the fire of our own guns and some from those of the rebels, shifting from side to side of the logs and trees as the fire came hotter from the one side or from the other, but always keeping up their own fire in the direction of the enemy, they maintained the unequal fight until an officer, sent for the purpose, succeeded in stopping the fire of our own guns, and the sharpshooters willingly withdrew from a position such as they had never found themselves in before, and from a scene which no man present will ever forget.

They were sharply pressed by the advancing enemy, but now, being out of the line of the enfilading fire from Hazel Grove, and no longer subject to the fire of their own friends, the withdrawal was made in perfect order, the line halting at intervals at the sound of the bugle and delivering well aimed volleys at the enemy, now fully exposed, and even at times making counter-charges to check their too rapid advance.

In one of these rallies there fell a man from another company whose death as well deserves to be remembered in song as that of the "sleeping sentinel." He had been condemned to death by the sentence of a court martial, and was in confinement awaiting the execution of the sentence when the army left camp at Falmouth at the outset of the campaign. In some manner he managed to escape from his guards, and joined his company on the evening of the second day's light. Of course it was irregular, and no precedent for it could possibly be found in the army regulations, but men were more valuable on that field than in the guard house; perhaps, too, his captain hoped that he might, in the furrow of the battle, realize his own expressed wish that he might meet his fate there instead of at the hands of a firing party of the provost guard, and thus, by an honourable death on the battle field, efface to some extent the stain on his character. However it was, a rifle was soon found for him (rifles without owners were plenty on that field), and he took his place in the ranks. During all of that long forenoon's fighting he was a marked man. All knew his history, and all watched to see him fall; for while others carefully availed themselves of such shelter as the field afforded, he alone stood erect and in full view of the enemy. Many times he exhausted the cartridges in his box, each time replenishing it from the boxes of his dead or wounded companions. He seemed to bear a charmed life; for, while death and wounds came to many who would have avoided either, the bullets passed him harmless by. At last, however, in one of the savage conflicts when the sharpshooters turned on the too closely following enemy, this gallant soldier, with two or three of his companions, came suddenly upon a small party of rebels who had outstripped their fellows in the ardour of the pursuit; he, being in the advance, rushed upon them, demanding their surrender. "Yes," said one, "we surrender," but at the same time, as —— lowered his gun, the treacherous rebel raised his, and the sharpshooter fell, shot through the heart. He spoke no word, but those who caught the last glimpse of his face, as they left him lying where he fell, knew that he had realized his highest hope and wish, and that he died content. The sequel to this sad personal history brings into tender recollection the memory of that last and noblest martyr to the cause of the Union, President Lincoln. The case was brought to his notice by those who felt that the stain upon the memory of this gallant, true hearted soldier was not fully effaced, even by his noble self-sacrifice, and would not be while the records on the books stood so black against him. The president was never appealed to in vain when it was possible for him to be merciful, and, sitting down, he wrote with his own hand a full and free pardon, dating it as of the morning of that eventful 3rd of May, and sent it to the widow of the dead soldier in a distant state. It was such acts as this that made Abraham Lincoln so loved by the soldiers of the Union. They respected the president, but Abraham Lincoln—the man—was *loved*.

Upon the arrival of the retreating riflemen at the new line in front of Fairview, they found their division, the main portion of which had, of course, preceded them, in line of battle in rear of the slight defences which had been thrown up at that point, where they enjoyed a brief period of much needed repose, if a short respite from actual personal encounter could be called repose. They were still under heavy artillery fire, while musketry was incessant and very heavy only a short distance away, the air above their heads being alive, at times, with everything that kills. Yet so great was their fatigue, and so quiet and restful their position in comparison with what it had been for so long a time, that, after receiving rations and a fresh supply of ammunition for their exhausted boxes, officers and men alike lay down on the ground, and most of them enjoyed an hour of refreshing sleep. So "*Use doth breed a habit in a man.*"

Their rest was not of long duration, however, for the rebels made a desperate and savage attack on the line in their front and the Third Corps soon found itself again engaged. The enemy, under cover of their artillery on the high ground at Hazel Grove, made an assault on what was now the front of the Union line, (if it could be said to have a front,) while the force which the sharpshooters had so long held in check during the early part of the day made a like attack on that line now the right of the entire army. So heavy was the attack, and so tenaciously sustained, that the Union troops were actually forced from their lines in front and on the flank of Fairview, and the hill was occupied by the rebels, who captured, and held for a time, all the Union guns on that eminence. It was at this stage of affairs that the Third Corps was again called into action, and charging the somewhat disorganized enemy they retook the hill with the captured guns, and following up the flying rebels, they drove them to, and beyond the position they had occupied in the morning. Here, however, meeting with a fresh line of the enemy and being brought to a check, they were ordered again to retire; for Hooker, by this time intent only upon getting his army safely back across the river, had formed still another new line near to, and covering, the bridges and fords by which alone could he place his forces in a position of even comparative safety. To this line then the Third Corps, with the tired and decimated sharpshooters, retired late in the afternoon, hoping and praying for a respite from their terrible labours. For a little time it looked, indeed, as if their hopes would be realized, but as darkness drew on the corps commander, desiring to occupy a wooded knoll at some little distance from his advanced picket line, and from which he anticipated danger, ordered Gen. Whipple, to whose division the sharpshooters had been returned, to send a brigade to occupy it. Gen. Whipple replied that he had one regiment who were alone equal to the task and to whom he would entrust it, and ordered the sharpshooters to attempt it.

Between this wooded hill and the position from which the regiment must charge was an open field about one hundred yards in width which was to be crossed under what might prove a destructive fire from troops already occupying the coveted position. It was a task requiring the most undaunted courage and desperate endeavour on the part of men who had already been for two full days and nights in the very face of the enemy, and they felt that the attempt might fairly have been assigned to a portion of the forty thousand men who, up to that time, had been held in reserve by Gen. Hooker for some inscrutable purpose, and who had not seen the face of an enemy, much less fired a shot at them; but they formed for the assault with cheerful alacrity. To Co. F was assigned the lead, and marching out into the open field they deployed as regularly as though on their old drill ground at camp of instruction. Corps, brigade and division commanders were looking on, and the men felt that now, if never before, they must show themselves worthy sons of the Green Mountain state. Led by their officers, they dashed out into the plain closely supported by the rest of the regiment. Night was rapidly coming on, and in the gathering gloom objects could hardly be distinguished at a distance of a hundred yards. Half the open space was crossed, and it seemed to the rushing men that their task was to be accomplished without serious obstructions, when, from the edge of the woods in front, came a close and severe volley betraying the presence of a rebel line of battle; how strong could only be judged by the firing, which was so heavy, however, as to indicate a force much larger than the attacking party. On went the brave men of Co. F, straight at their work, and behind them closely followed the supporting force. In this order they reached the edge of the forest when the enemy, undoubtedly supposing from the confidence with which the sharpshooters advanced that the force was much larger than it really was, broke and fled and the position was won.

From prisoners and wounded rebels captured in that night attack it was learned that the force which had thus been beaten out of a strong position by this handful of men was a portion of the famous Stonewall brigade, Jackson's earliest command, and they asserted that it was the first time in the history of the brigade that it had ever been driven from a chosen position. The sharpshooters were justly elated at their success and the more so when Gen. Whipple, riding over to the point so gallantly won, gave them unstinted praise for their gallant action. In this affair the regiment lost many gallant officers and men, among whom were Lieut. Brewer of Co. C and Capt. Chase, killed, and Major Hastings and Adjt. Horton, wounded. In Co. F Michael Cunningham, J. S. Bailey and E. M. Hosmer were wounded.

Major Hastings had not been a popular officer with the command. Although a brave and capable man, he was of a nervous temperament, and in the small details of camp discipline was apt to be over zealous at times. He had, therefore, incurred the dislike of many men, who were wont to apply various opprobrious epithets to him at such times and under such circumstances as made it extremely unpleasant for him. Such were the methods adopted by some soldiers to make it comfortable for officers to whom they had a dislike.

In the case of the major, however, this was a thing of the past. On this bloody field the men learned to respect their officer, and he, as he was borne from the field, freely forgave the boys all the trouble and annoyance they had caused him, in consideration of their gallant bearing on that day. Adjt. Horton, also a brave and efficient officer, received a severe wound—which afterwards cost him his good right arm—while using the rifle of J. S. Bailey of Co. F, who had been wounded.

Co. F, which, it will be remembered, had been acting as skirmishers, were pushed forward in advance of the main portion of the regiment to further observe the movements of the enemy and to guard against a surprise, and shortly afterwards were moved by the flank some two hundred yards to the right, and were soon after relieved by a force of infantry of the line which had been sent up for that purpose. While retiring toward the position to which they were directed, they passed nearly over the same ground which they had just vacated when they moved by the right flank, as previously mentioned, and received from the concealed rebels, who had reoccupied the line, a severe volley at close range. Facing to the right, Co. F at once charged this new enemy and drove them in confusion from the field. Lying down in this advanced position they passed the remainder of the night in watchful suspense.

At day break on the fourth day of the battle, Co. F was relieved from its position on the picket line and returned to the regiment, which was deployed as skirmishers, and led the van of Whipple's division in a charge to check movements of the enemy which had for their apparent object the interposition of a rebel force between the right wing of the army and its bridges. Firing rapidly as they advanced, and supported by the division close on their heels, they drove the enemy from their rifle pits, which were occupied by the infantry of the Third Corps, the sharpshooters being still in front. Here they remained, exchanging occasional shots with the rebel sharpshooters as occasion offered, for some hours. Hooker was not minded to force the fighting at Chancellorsville; preferring to await the result of Sedgwick's battle at Salem Church, which had raged furiously on the preceding afternoon until darkness put an end to the strife, and the tell tale guns of which even now gave notice of further effort.

Lee, however, pugnacious and aggressive, determined to renew his attack on the right, and, if possible, secure the roads to the fords and bridges by which alone could the defeated army regain the north bank of the river. With this view he reinforced Jackson's (now Stuart's) corps, and organized a powerful attack on the position of the Third Corps. The force of the first onset fell on the sharpshooters, who fought with their accustomed gallantry, but were forced by the weight of numbers back to the main line. Here the fighting was severe and continuous. The one party fighting for a decisive victory, and the other, alas, only bent on keeping secure its last and only line of retreat; but the incentive, poor as it was, was sufficient, and the rebels were unable to break the line. After four hours of continued effort they abandoned the assault and quiet once more prevailed. In this fight Gen. Whipple, the division commander, was killed. He was a gallant and an able soldier, greatly beloved by his men for the kindliness of his disposition. He had an especial liking for and confidence in the sharpshooters, which was fully understood and appreciated by them, and they felt his death as a personal loss.

To add to the horrors of this bloody field, on which lay nearly nine thousand dead and wounded Union soldiers and nearly or quite as many rebels, the woods took fire and hundreds of badly wounded men, unable to help themselves, and hopeless of succour, perished miserably in the fierce flames. Nothing in the whole history of the war is more horrible than the recollection of those gallant men, who had been stricken down by rebel bullets, roasted to death in the very presence of their comrades, impotent to give them aid in their dire distress and agony.

Oh, happy dead who early fell.

It was reserved for the *wounded* to experience the agonies of a ten-fold death. Hour after hour the conflagration raged, until a merciful rain quenched it and put an end to the horrible scene. The Third Corps remained in their position during the night, the sharpshooters, oddly enough as it seemed to them, with a strong line of infantry behind works between them and the enemy. Nothing occurred to break their repose, and for the first time for seven days they enjoyed eight hours of solid sleep unbroken by rebel alarms.

At day break on the morning of the 5th of May they were aroused by the usual command of "sharpshooters to the front," and again found themselves on the picket line confronting the enemy. The day passed, however, without serious fighting, one or two attacks being made by rebel skirmishers, more, apparently, to ascertain if the Union troops were actually there than for any more serious business.

These advances were easily repulsed by the sharpshooters without other aid, and at nine o'clock p.m., after seventeen hours of continuous duty without rations—for the eight days rations with which they started from their camp at Falmouth had long since been exhausted, and the scanty supply they had received on the afternoon of the third was barely enough for one meal—they were relieved and retired to the main line. The company lost on this day but one man, Martin C. Laffie, shot through the hand. Laffie was permanently disabled by his wound, and on the 1st of the following August was transferred to the Invalid Corps and never rejoined the company. Several prisoners were captured by the men of Co. F on that day, but on the whole it was, as compared with the days of the preceding week, uneventful. On the 6th the army re-crossed the Rappahannock by the bridges which had been preserved by the stubborn courage of the Third Corps, and the battle of Chancellorsville passed into history.

The sharpshooters returned to their old camp at Falmouth as they had returned to the same camp after the disastrous battle of Fredericksburgh. It seemed as though they were fated never to leave that ground to fight a successful battle. Only eight days before they had marched out with buoyant anticipations, full of courage and full of hope. They returned discouraged and dispirited beyond description.

At Fredericksburgh the army had marched to the attack without hope or expectation of victory, for their soldiers' instinct told them that that was impossible. At Chancellorsville, however, they felt that they had everything to hope for—a magnificent army in full health and high spirits, an able and gallant commander, for such he had always shown himself to be, and a fair field. The thickets of the wilderness, it is true, were dense and well nigh impassable for them, but they were as bad for the enemy as for themselves, and they had felt that on anything like a fair field they ought to win. Now they found themselves just where they started; they had left seventeen thousand of their comrades dead, or worse than dead, on the field, and fourteen guns remained in the hands of the rebels as trophies of their victory; guns, too, that were sure to be turned against the federals in the very next battle. Twenty thousand stand of small arms were also left on the field to be gathered up by the victors. It was a disheartening reflection, but soldier-like the men put it from their thoughts and turned their minds and hands to the duties and occupations of the present. In this battle Co. F lost Edward Trask, Jacob S. Bailey, Almon D. Griffin, Martin C. Laffie and John Monahan, wounded, besides several more whose names do not now occur to the writer. Bailey had been previously wounded at Malvern Hill and on this occasion his wound necessitated the amputation of his left arm, and he was honourably discharged from the service on the twenty-sixth of the following August. Monahan was transferred to the Invalid Corps and Griffin returned to his company and remained with it to be honourably mustered out by reason of expiration of term of service, on the 13th of September, 1864. Trask returned to his company to serve with it until the 5th of May, 1864, when he was killed in the battle of the Wilderness.

CHAPTER 5

Gettysburg to the Wilderness

From the date of their return from the field of Chancellorsville to the 11th of June, the sharpshooters remained in camp
near Falmouth engaged only in the usual routine duties of camp
life. Drills, reviews and other parades of ceremony were of frequent occurrence, but nothing of moment took place to essentially vary the monotony of their lives. Occasionally a detail
would be made from the company for a day or two of especial
service at some portion of the picket line where the rebel sharpshooters had become unusually aggressive, but affairs in those
parts generally soon became satisfactory, and the men would be
ordered back to camp. These little episodes were eagerly wel-
comed by men tired again of the inactivity of their lives in permanent camp. During this time, however, important changes in
the organization of the company took place. Capt. Seaton, who
had never entirely recovered from the effects of his wound received at Malvern Hill, resigned on the 15th day of May, and E.
W. Hindes was appointed and commissioned captain. C. D. Merriman was promoted to be first lieutenant and H. E. Kinsman
second lieutenant, the two former to date from May 15, 1863,
and the latter from May 26.

The non-commissioned officers were advanced to rank as
follows:

First Sergeant	Lewis J. Allen.
Second Sergeant	A. H. Cooper.
Third Sergeant	Cassius Peck.

Fourth Sergeant	Paul M. Thompson.
Fifth Sergeant	Edward F. Stevens.
First Corporal	Jacob S. Bailey.
Second Corporal	L. D. Grover.
Third Corporal	Chas. M. Jordan.
Fourth Corporal	E. M. Hosmer.
Fifth Corporal	Edward Trask.
Sixth Corporal	W. H. Leach.
Seventh Corporal	M. Cunningham.
Eighth Corporal	Edward Lyman.

The new officers had been connected with the company from its organization; they were all roll of honour men, straight up from the ranks, and were men of distinguished courage and skill, as they had demonstrated already on at least fifteen occasions upon which the Army of the Potomac had been engaged in pitched battles with the enemy, besides numberless minor engagements and skirmishes. Indeed, their lives might be said to have been passed, for the year and a half they had been in the field, in constant battle, and the same was true of every man in the company as well. The month of June was, however, destined to bring with it hard marches and stirring events.

Not content with the results of the Maryland campaign of 1862, which had resulted in a disastrous rebel defeat at Antietam, Lee, perhaps recognizing the historical fact that a power which allows itself to be placed entirely on the defensive is sure to be beaten in the end, determined to essay once more an invasion of the loyal states, and to transfer the seat of war, if possible, from the impoverished and suffering South, to the soil of populous and wealthy Pennsylvania.

His route was substantially the same one pursued by him the previous year, but not now, as on that occasion, was the severe fighting to take place on the soil of Virginia.

By skilful feints and rapid marches, he succeeded in placing his army north of the Potomac before the Union commander could strike a blow at him. Early in the month it was certain that Lee was about to take the field in some direction. Sick and wounded were sent to northern hospitals, all surplus baggage and stores were turned in, and the Union army, stripped of everything but what the men carried on their persons, was ready to follow or to confront him. On the 11th of June the sharpshooters broke camp at five o'clock p.m., and, for the third time, marched out from the ground that had been their home for nearly seven months. Twice before had they left the same place to fight desperate battles with the same enemy, and twice had they returned to it, defeated and despondent. Many a man, as the regiment marched out, wondered in his heart if such would be their fate again; but soldiers are optimists by nature and education; they soon learn that to fear and dread defeat is to invite it; that confidence begets confidence, and that the example of courage and cheerfulness is contagious. Not for a long time, therefore, did these gloomy thoughts possess their minds, and soon they were stepping out merrily to the sound of the bugle.

Other portions of the army had preceded them, and still others were starting by different roads; and as far as the eye could reach, as the columns passed over some height of land, could be seen the clouds of dust that, rising high in the air, betrayed the presence of marching men. Pressing rapidly northward, passing successively Hartwood church, Rappahannock Station, Catlet's Station, Manassas Junction, Centerville and Green Springs—all familiar as the scenes of past experience, and many of them sacred to the memory of dead comrades— they forded the Potomac at Edwards' Ferry on the 25th of June and reached the mouth of the Monocacy, having marched thirty-one miles on that day. Arriving at that point, tired and foot-sore, as may be imagined after such a march, they found an *aide-de-camp* ordered to conduct them to their allotted camp ground. He appeared to be one of those nice young men who were so often appointed to positions on the staff for their beauty or their fragrance, or for the general elegance of manners, rather than for their ability to be of any real service. This young person, with no apparent idea of where he wanted to go, marched them up and down and around and about, until the patience of Trepp, the Dutch lieutenant-colonel, was exhausted. Commanding halt, he turned to the bewildered aide and with phrases and objurgations not fitted for the polite ears of those who will read this book, concluded his lecture with "Now mine frent, dese men is tired and dey is to march no more dis day," then, turning to the regiment, he commanded, in tones that might have been heard at Washington, "Men, lie down!" and the sharpshooters camped just there.

Leaving this place on the twenty-sixth, they marched to Point of Rocks, and on the twenty-seventh to Middletown. On the twenty-eighth they marched via Frederick and Walkersville and crossed the Catoctin Mountains at Turner Gap. On this day the corps commander, General Sickles, returned to his command after a short absence, and on the same day General Hooker, not being able to make his ideas of the campaign square with those of the department generals at Washington, was relieved, at his own request, and General Meade was appointed to the command. The army parted with Hooker without very much regret. They recognized his wonderful fighting qualities as a division or corps commander, and he was personally popular, but they had never quite forgiven him for Chancellorsville, where he took his army, beaten and well nigh crushed, back from an enemy numerically weaker than his own, while he had yet nearly forty thousand soldiers who had not been engaged in the action, and hardly under fire. It is safe to say that his army had no longer that degree of confidence in his ability to handle large armies, and to direct great battles, so essential to success. Of his successor the army only knew that he was a scholarly, polished gentleman, personally brave, and that as a brigade, division and corps commander he had made few mistakes. On the whole, his record was favourable and the men marched willingly under him, although the choice of the rank and file might possibly have been some other man.

On the twenty-ninth the sharpshooters marched with the corps to Taneytown, some twenty miles distant, and on the next day to within two miles of Emmetsburg, where they camped for the night. On the morning of July 1st the guns of Reynold's fight at Gettysburg were plainly heard, and in the late afternoon they started for the point of action, some ten miles distant, making most of the distance at the double quick.

At about sunset they arrived on the field and went into bivouac in the rear of the hill known in the history of the subsequent battle as Little Round Top, and were once more confronting their ancient antagonists. The sharpshooters were now attached to the second brigade, commanded by Gen. J. H. H. Ward, of the first division, under Gen. Birney, the old third division having been consolidated with the first and second after the terrible losses of the corps at Chancellorsville, and in this connection we shall have to follow them through the battle of Gettysburg. The battle of the 1st of July was over. The First and Eleventh Corps had sustained a serious defeat, and at the close of that day the rolls of these two corps showed the terrible loss of over nine thousand men, and yet the battle had hardly commenced. The situation was not an encouraging one to contemplate; not half the Union army was up, some corps being yet thirty or forty miles distant, while the events of the day showed that the rebel army was well concentrated—but the die was cast, events forced the battle then and there, and thus the rocky ridges of Gettysburg became of historic interest and will remain so forever.

Troops arrived rapidly during the night and were assigned places, as they arrived, in the chosen line, which was in a direction nearly north and south. The extreme left rested on a rocky height rising some three hundred feet above the level of the surrounding country; some five hundred yards to the north of this hill, called Round Top, rises a similar elevation, although of less height, called Little Round Top; thence north to Cemetery Hill, immediately overlooking the village of Gettysburg about two miles distant, the Union troops occupied, or were intended to occupy, a rocky ridge overlooking and commanding the plain to the westward. From Cemetery Hill the line was refused and curved backward to the east until the extreme right rested on a wooded eminence called Culp Hill, and fronted to the east, so that the entire line was some three miles, or perhaps a little more, long, and was in shape like a fish hook, the shank lying along the ridge between Round Top and Cemetery Hill, and the point on Culp Hill. Below the bend of the hook, at the base of Cemetery Hill, lay the village of Gettysburg. Such was the Union position at daylight on the morning of the 2nd of July, 1863. Fronting that portion of the federal troops which was faced to the west, and at a distance of about one mile, ran another ridge, parallel to the first, called Seminary Ridge, and which was occupied by the Confederate army. To the north and east of Gettysburg the ground was open, no ridges or considerable body of wood land existed to cover or screen the movements of the rebel troops. The village of Gettysburg was occupied by the enemy on the afternoon of the 1st of July after the defeat of the First and Eleventh Corps, and yet remained in their possession. Midway between the two armies ran the Emmetsburg road, following the crest of a slight elevation between the two lines of battle. The position assigned to the Third Corps was that portion of the line immediately north of Little Round Top where the ridge is less high than at any other portion. Indeed, it sinks away at that point until it is hardly higher than the plain in front, and not as high as the ridge along which runs the Emmetsburg road. At an early hour on the morning of the 2nd, Sickles, believing himself that the latter ridge afforded the better position, and perhaps mistaking Gen. Meade's instructions, passed down into the valley and took up the line of the Emmetsburg road, his centre resting at a point known in the history of the battle as the "peach orchard." From this point his line was prolonged to the right by Humphrey's division along the road, while Birney's division, to which Ward's brigade with the sharpshooters was attached, formed the left, which was refused; the angle being at the peach orchard, and the extreme left resting nearly at the base of Round Top, at a point known by the altogether suggestive and appropriate name of the Devil's Den—a name well applied, for a more desolate, ghostly place, or one more suggestive of the home of evil spirits can hardly be imagined. Barren of tree or shrub, and almost destitute of any green thing, it seems cursed of God and abandoned of man.

Pending the deployment of the Third Corps, four companies of the sharpshooters, F, I, D and E, with the Third Maine, a small regiment of only two hundred men, were detached from Ward's brigade and ordered to a point in front and to the right of the peach orchard, where they were directed to advance to a piece of wooded land on the west of the Emmetsburg road and feel for the enemy at that point. The four companies, deployed as skirmishers, advanced in a north-westerly direction, and at about nine o'clock encountered a strong force of the rebels, consisting of at least one brigade of Longstreet's command, who, with arms stacked, were busily engaged in preparing their breakfast when the rifles of the sharpshooters gave them notice of other employment. They were taken entirely by surprise, and quickly perceiving this fact, the riflemen dashed forward, firing as they pressed on as rapidly as the breech loaders could be made to work. The rebels made but a short stand; taken entirely unprepared and unaware of the insignificant numbers of the oncoming force, they seized their guns from the stack, and, after one or two feeble volleys, retreated in confusion.

The general in command made a gallant personal effort to rally his men, but fell dead from his horse immediately in front of Co. F. The rout of the enemy at this point was now complete, and pressing their advantage to the utmost the sharpshooters drove them back nearly to the main rebel line on Seminary Ridge, capturing many prisoners who were sent to the rear, and a large number of small arms which, however, they were unable to bring away. Having thus cleared the ground nearly to the main rebel line, they took position behind walls, fences, etc., and for the two or three hours following were engaged in sharpshooting with the enemy similarly posted in their front. Their position was now some distance to the right of the peach orchard and in front of the right, or right centre, of Humphrey's division.

At about half-past three in the afternoon Longstreet commenced his attack on Sickles' extreme left near Round Top, the battle soon becoming very severe also at the angle in the peach orchard and involving Humphrey further to the right. The attacking columns had passed to the left of the sharpshooters and the fighting was now in their left and rear. The rebels in their front also became very aggressive and they were gradually pushed back until they became intermingled with the troops of Humphrey's division posted along the Emmetsburg road where the struggle soon became close and deadly. The angle at the peach orchard was the key to Sickles' line, and against it Longstreet pushed his best troops in dense masses, and at this point occurred some of the hardest fighting that took place on the whole field; but as the troops whose doings are chronicled in these pages had no part in that struggle, it is enough to say that after a gallant resistance the line was broken at the angle and the shouting rebels, rushing through the gap, took both portions of the line in reverse, while both portions were yet resisting heavy attacks on their fronts. Such a situation could have but one result—both wings were compelled to retire in confusion.

Anticipating this, Meade had ordered heavy supporting columns to be formed behind the crest of the ridge and these were ordered down to the relief of the sorely tried Third Corps. Barnes' division of the Fifth Corps, the same to which the sharpshooters had been attached for so long a time, and in the ranks of which they had fought in all the battles previous to Fredericksburgh, came gallantly to the rescue, but were unable to withstand the terrible vigour of the Confederate assault, and Caldwell's division of the Second Corps was also thrown in to check the onset.

These troops fought with the greatest courage but were defeated with the loss of half the men engaged. In the mean time Longstreet, finding the ground between the left of Birney's division and the base of Round Top unoccupied, pushed a force behind the Union left at that point which succeeded in gaining a position in the rocky ravine between the two Round Tops from which they pushed forward to secure the possession of the lesser elevation, at that moment unguarded. This was the key to the entire Union line, and once in the hands of the rebels would probably decide the battle in their favour. But Warren, another old Fifth Corps friend, quickly discovered the danger and ordered Vincent with his brigade to occupy and defend this important point. The struggle for its possession was terrible, but victory perched upon the Union banners and the hill was made secure. Vincent and Hazlett, both of the Fifth Corps also, were killed here. They had been well known and highly esteemed by many of the officers and men of the sharpshooters, and by none were they more sincerely lamented.

Darkness put an end to the battle of July 2nd. Lee had gained considerable ground, for the whole of the line occupied by the Third Corps was now in his possession. There yet remained for him to carry the real line of the federal defences which was as yet intact. The position taken by Gen. Sickles had been intrinsically false, and was one from which he would have been withdrawn without fighting had time allowed. Lee had gained ground, and that was all, unless the inspiriting effects of even partial success can be counted.

Many thousands of Union soldiers lay dead and wounded on the field, and the Army of the Potomac was the weaker by that number of men, but Lee had lost an equal, or more likely a greater number, so that on the whole the result of the day could not be counted as a substantial gain for the rebels, and when the federals lay down for the night, it was with confidence and assurance that the morrow would bring its reward for the mishaps of the day. The corps commander, Gen. Sickles, had been wounded and Gen. Birney succeeded to the command. Gen. Ward took command of the division, and thus it came about that Col. Berdan was in command of the brigade.

Company F had killed on this day Sergeant A. H. Cooper, and Geo. Woolly and W. H. Leach wounded. Woolly's wound was severe and resulted in the loss of his arm. Other companies in the regiment had suffered more or less severely, the four companies engaged in front and to the right of the peach orchard losing twenty men, killed and wounded, out of the one hundred engaged.

During the night succeeding the 2nd of July the shattered remains of the Third Corps was withdrawn from the front line and massed behind the sheltering ridge as a reserve. Its terrible losses of the day, added to those sustained at Chancellorsville, had reduced the once powerful corps almost to the proportions of a brigade. As the troops stood in line the colours were like a fringe along its front, so close together were they. The regiments that defended them were like companies—indeed, many regiments had not the full number of one hundred men which is called for on paper by a full company. The Third Corps was nearly a matter of history, but the few men left with their colours were veterans, tried and true, and although they were not displeased to be relieved from the active fighting yet in store for the federals, they were quite ready to stand to arms again whenever it should please Gen. Meade to so direct. At daylight the enemy opened a heavy artillery fire all along the line. The random nature of the firing was proof, however, that nothing more serious than demonstration was intended.

Late at night on the preceding day the rebels had succeeded in gaining important ground on the extreme right, and had indeed possessed themselves of almost the whole of the wooded eminence known as Culp's Hill, from which their artillery, should they be allowed time to get it up, would take almost the entire Union line in the rear. To regain this, Geary's division was sent in early in the day, and after four hours of severe fighting the rebels were dislodged and the Union right was restored. Affairs now became quiet and so remained for some hours—suspiciously quiet indeed, and all felt that some great effort was about to be made by the Confederates. At about one o'clock a single gun was fired as a signal from the Confederate lines near the seminary, and instantly one hundred and fifteen guns opened on the Union centre, which was held by the First and Second Corps, supported by all that remained of the Third. Never before had the Union troops been subjected to such an artillery fire. Previous to this battle the cannonading at Malvern Hill had always been quoted as the heaviest of the war. The bombardment of Fredericksburgh had also been on a magnificent scale, but here the troops were to learn that still further possibilities existed. Eighty Union guns responded vigorously, and for two hours these guns—nearly two hundred in number—hurled their shot and shell across the intervening plain in countless numbers. The Union artillery was posted along the crest of, or just behind the ridge, while the lines of infantry were below them on the western slope. The soldiers lay prone on the ground, sheltering themselves behind such inequalities of the surface as they could find, well knowing that this awful pounding was only the precursor of a struggle at closer quarters, which, if less demonstrative and noisy, would be more deadly; for experience had taught them that however frightful to look at and listen to, the fire of shell at such long range was not, on the whole, a thing to inspire great fear. It is a curious fact, however, that heavy artillery fire, long sustained, begets an irresistible desire to sleep; and hundreds of Union soldiers went quietly to sleep and slept soundly under the soothing influence of this tremendous lullaby.

At three o'clock the artillery fire ceased, and from the woods crowning Seminary Ridge, a mile away, swarmed the grey coated rebels for another attempt on the federal line. Lee had tried the left and had failed; he had been partially successful on the right on the preceding evening, but had been driven back in the morning. It only remained for him to try the centre. In the van of the charging column came Picket's division of Virginia troops, the flower of Lee's army, fresh and eager for the strife. On his right was Wilcox's brigade of Hill's corps, and on his left Pender's division. Could Picket but succeed in piercing the Union centre, these two supporting columns, striking the line at points already shattered and disorganized by the passage of Picket's command, might be expected to give way in turn, and the right and left wings of the federal army would be hopelessly separated. But others besides Lee saw this, and Meade hastened to support the points on which the coming storm must burst with all the troops at his command. The Third Corps was ordered up and took position on the left of the First, directly opposite the point at which Wilcox must strike the line, if he reached so far. Our artillery, which had been nearly silent for some time, opened on the oncoming masses as they reached the Emmetsburg road with canister and case shot which made fearful gaps in their front, but closing steadily on their colours they continued to advance. Their courage was magnificent and worthy of a better cause. Eight Union batteries, brought forward for the purpose, poured an enfilading fire into the rushing mass, while Stannard's Second Vermont Brigade, far in advance of the main line, suddenly rose up and, quickly changing front, forward on the right, commenced a close and deadly fire directly on their exposed right flank. Their track over that open plain was marked by a swath of dead and dying men as wide as the front of their column; still they struggled on and some portion of the attacking force actually pierced the Union line, and the rebel Gen. Armistead was killed with his hand upon one of the guns of Wheeler's battery. The point had been well covered, however, and no sooner did the rebel standards appear crowning the stone wall, which was the principal defensive work, than the troops of the second line were ordered forward and for a few moments were engaged in a fierce hand to hand fight over the wall. The force of the rebel attack was, however, spent; exhausted by their march of a mile across the plain in the face of the deadly fire, and with ranks sadly thinned, the rebels, brave as they undoubtedly were, were in no shape to long continue the struggle. They soon broke and fled, thousands, however, throwing down their arms and surrendering themselves as prisoners rather than risk the dangerous passage back to their own lines, a passage only in a degree less perilous than the advance.

In the meantime Wilcox, on the right, had pushed gallantly forward to strike the front of the Third Corps where the sharpshooters had been posted in advantageous positions to receive him. They had opened fire when he was some four hundred yards away, too far for really fine shooting at individual men, but not so far as to prevent considerable execution being done on the dense masses of men coming on. This attack, however, was not destined to meet with even the small measure of success which had attended Picket's assault, for Col. W. G. Veazey of the Sixteenth Vermont, one of the regiments of Stannard's Second Vermont Brigade, which had been thrown forward on the right flank of Picket's column, seeing that attack repulsed, and being aware of the approach of Wilcox in his rear, suddenly counter-marched his regiment and made a ferocious charge on the left of Wilcox's column, even as he had just done on the right of Picket's. The effect was instantaneous; they faltered, halted, and finally broke. Launching forward, Veazey captured many prisoners and colours, many more, in fact, than he had men in his own ranks.

The fighting of the 3rd of July now ceased and the federals had been signally successful. The morrow was the 4th of July, the birthday of the nation; would it be ever after celebrated as the anniversary of the decisive and closing battle of the war? Many hearts beat high at the thought, and the troops lay on their arms that night full of hope that the end was at hand.

The repulse of Lee's final assault on the 3rd of July had been so complete and crushing, so apparent to every man on the field, that there were none who did not awake on the morning of the 4th with the full expectation that the Army of the Potomac would at once assume the offensive and turn the repulse of the last two days into such a defeat as should insure the utter destruction of the rebel army. Everything seemed propitious; Sedgwick's gallant Sixth Corps had arrived late on the night of the second, and had not been engaged. The men were fresh and eager to deliver on the national holiday the death blow to the rebellion. The troops who had been engaged during that terrible three days battle were equally eager, notwithstanding their labours and sufferings, but Meade was eminently a conservative leader, and feared to

Put it to the touch
To win or lose it all.

And so the day was spent in such quiet and rest as could be obtained by the men. The wounded were gathered and cared for, rations and ammunition were issued, and every preparation for further defence should Lee again attack, or for pursuit should he retreat, was made. Some rather feeble demonstrations were made at various points, but no fighting of a serious character took place on that day. The sharpshooters were thrown forward as far as the peach orchard where they took up a position which they held during the day, constantly engaged in exchanging shots with the rebel pickets posted behind the walls and fences in the open field in front of the woods behind which lay the rebel army. It was of itself exciting and dangerous employment; but, as compared with their experiences on the two preceding days, the day was uneventful. Co. F lost here, however, two of its faithful soldiers, wounded, L. B. Grover and Chas. B. Mead. Both recovered and returned to the company, Grover to be promoted sergeant for his gallantry on this field, and Mead to die by a rebel bullet in the trenches at Petersburg. The regiment as a whole had suffered severely. The faithful surgeon, Dr. Brennan, had been severely wounded while in the discharge of his duty in caring for the wounded on the field, and Capt. McLean of Co. D was killed.

Many others, whose names have been lost in the lapse of years, fell on this bloody field. The fifth was spent in gathering the wounded and burying the dead. On the sixth Meade commenced that dilatory pursuit which has been so severely criticised, and on the twelfth came up with the rebel army at Williamsport, where Lee had taken up and fortified a strong position to await the falling of the river, a sudden rise of which had carried away the bridges and rendered the fords impassable.

The army was eager to attack; flushed with their success, and fully confident of their ability to give rebellion its death blow, they fairly chafed at the delay—but Meade favoured the cautious policy, and spent the twelfth and thirteenth in reconnoitring Lee's position. Having finished this preliminary work, he resolved on an attack on the fourteenth; but Lee, having completed his bridges, made a successful passage of the river, and by eight o'clock on that morning had his army, with its trains and stores, safe on the Virginia side.

On the seventeenth the Third Corps crossed the river at Harper's Ferry and were once more following a defeated and flying enemy up the valley, over the same route by which they had pursued the same foe a year before while flying from Antietam. The pursuit was not vigorous—the men marched leisurely, making frequent halts. It was in the height of the blackberry season, and the fields were full of the most delicious specimens. The men enjoyed them immensely, and, on a diet composed largely of this fruit, the health of the men improved rapidly.

On the nineteenth the sharpshooters reached Snicker's Gap, where, on the 3rd of the previous November, they had looked down on the beautiful valley of Virginia and beheld from their lofty perch Lee's retreating columns marching southward. To-day, from the same point of view, they beheld the same scene; but how many changes had taken place in that little company since they were last on this ground! Death, by bullet and by disease, had made sad inroads among them, and of the whole number present for duty the previous November, less than one-half were with their colours now, the others were either dead in battle, or of wounds received in action, or honourably discharged by reason of disability incurred in the service. Sheridan once said that no regiment was fit for the field until one-half of its original numbers had died of disease, one-quarter been killed in action, and the rest so sick of the whole business that they would rather die than live. Judged by this rather severe standard, Co. F was now fit to take rank as veterans. Descending the mountains, they marched southward, passing the little village of Upperville on the twentieth.

On the twenty-third the Third Corps was ordered to feel the enemy at Manassas Gap, and there ensued a severe skirmish, known as the affair of Wapping Heights. The sharpshooters opened the engagement and, indeed, bore the brunt of it, dislodging the enemy and driving them through the gap and beyond the mountain range. They inflicted considerable loss on the rebels, and made a number of prisoners.

In this affair a man from another company came suddenly face to face with an armed rebel at very short range; each, as it subsequently appeared, had but one cartridge and that was in his gun. Each raised his rifle at the first sight of the other and the reports were simultaneous. Both missed—the rebel bullet struck a tree so close to the sharpshooter's face that the flying fragments of bark drew blood; the Union bullet passed through the breast of the rebel's coat, cutting in two in its passage a small mirror in his breast pocket. They were now upon equal terms but each supposed himself at the disadvantage. Yankee cheek was too much, however, for the innocent Johnnie, for the sharp-shooter, with great show of reloading his rifle, advanced on the rebel demanding his surrender. He threw down his gun with bad grace, saying as he did so:

"If I had another cartridge I would never surrender."

"All right, Johnnie," said the Yankee, "If I had another you may be sure I would not ask you to surrender."

But Johnnie came in a prisoner. In this action the sharpshoot-ers expended the full complement of sixty rounds of ammuni-tion per man, thus verifying the assertion of their ancient enemy in the ordnance department that "the breech loaders would use up ammunition at an alarming rate;" both he and others were by this time forced to admit, however, that the ammunition was expended to very useful purpose. Passing now to the south-east over familiar grounds they encamped at Warrenton on the twenty-sixth, and on the thirty-first at or near White Sulphur Springs, where they remained until the 15th of September, en-joying a much needed rest. It was eighty-one days since they left their camp at Falmouth to follow and defeat Lee's plans for an invasion of the North, and during that time they had not had one single day of uninterrupted rest. Here the regiment had the first dress parade since the campaign opened.

On the 15th of September they broke camp and marched to Culpepper, some ten miles to the southward, where they remained until the 10th of October. On the 22nd of September eight days rations had been issued and it looked as though serious movements were contemplated, but the plan, if there was one, was not carried out.

On the 11th of October, with full haversacks and cartridge boxes, they broke camp and moved again northward, crossing the Rappahannock by Freeman's ford, near which they remained during the rest of that day and the whole of the twelfth on the picket line, frequently engaged in unimportant skirmishes with the enemy's cavalry. On the thirteenth they marched in the early morning, still towards the north, prepared for action, and at Cedar Run, a small tributary of the Rappahannock, they found the enemy in considerable force to dispute the crossing. Here a severe action took place, and as the emergency was one which did not admit of delay, the attack was made without the formality of throwing out skirmishers, and the sharpshooters charged with the other regiments of the division in line of battle. Edward Jackson was severely wounded here, but returned to his company to remain with it to the close of the war. Quickly brushing away this force the corps advanced northwardly by roads lying to the west of the Orange & Alexandria railroad and parallel with it, and after a fatiguing march arrived at Centerville, only a few miles from Washington.

The cause of this rapid retrograde movement was not easily understood by the men at the time, but was subsequently easily explained. Lee had not been satisfied with the results of his three previous attempts to destroy the Union army by turning its right and cutting it off from Washington, and had essayed a fourth. It had been a close race, but the Union commander had extricated his army from a position that, at one time, was one of grave peril, and had it compact and ready on the heights of Centerville with the fortifications of Washington at his back. Lee was now far from his own base of supplies and must attack the Union army in position at once, or retreat. He took one look at the situation and chose the latter alternative, and on the nineteenth the Army of the Potomac was once more in pursuit, the Third. Corps with the sharpshooters passing Bristoe's Station on that day with their faces toward the South. On the twentieth they forded Cedar Run at the scene of their battle of the week before, and on the same day, owing to an error by which the sharpshooters were directed by a wrong road, they re-crossed it to the north bank, from which they had, later in the day, to again ford it to reach their designated camping place on the south side near Greenwich, thus making three times in all that they waded the stream on this cold October day, sometimes in water waist deep. The next camp made was at Catlet's Station, when the sharpshooters with the Third Corps remained inactive until the 7th of November awaiting the repairing and reopening of the Orange & Alexandria railroad which had been greatly damaged by Lee in his retreat, and which, as it was the main line of supply for Meade's army, it was necessary to repair before the army could move further southward.

On the seventh, the railroad having been completely repaired and the army fully supplied with rations, ammunition and other necessary articles, Meade determined to try to bring his enemy to a decisive action in the open field, and to that end directed the right wing of his army, consisting of the Fifth and Sixth Corps under Sedgwick, to force the passage of the Rappahannock at Rappahannock Station, while the left wing, consisting of the First, Second and Third Corps, was directed on Kelly's Ford, some five miles lower down the river.

The Third Corps, under Birney, had the advance of the column, the sharpshooters acting as flankers, until the head of the column arrived at the river opposite the designated crossing place. The enemy were found in strong force occupying rifle pits on the opposite bank, and the column was deployed to meet the exigency of the occasion. The sharpshooters were at the front as skirmishers and advanced at the double quick in splendid order until they reached the bank of the river, when they took such cover as was afforded by the inequalities of the ground, and commenced an active fire upon the enemy in the rifle pits on the opposite side. It was soon found, however, that they could not be driven from their strong position by simple rifle work, and the regiment was ordered to cross the stream and drive them out by close and vigorous attack. It was not a cheerful prospect for the men who were to wade the open stream nearly waist deep and exposed to the cool fire of the concealed enemy, who would not aim less coolly because the sharpshooters would necessarily be unable to return the fire; but the line was carefully prepared and at the sound of the bugle every man dashed forward into the cold and rapid water and struggled on. Co. F was one of the reserve companies and thus followed the skirmishers in column of fours instead of in a deployed line. As the skirmishers arrived on the further shore they naturally took such cover as they could get, and opened a rapid fire. The Vermonters, however, closely following the movement, passed the skirmish line thus halted and pushed on without stopping to deploy even. Capt. Merriman, who had just succeeded to his well deserved promotion, led the way until he stood upon the very edge of the works overlooking the rebels within, of whom he demanded an immediate and unconditional surrender. He was far in advance of his men, and the rebels, at first taken aback by the very boldness of the demand, now seeing him unsupported as they thought, refused with strong language to surrender, but on the contrary called upon him to yield himself up as their prisoner. Merriman, however, was not minded to give up his captain's sword on the very first day he had worn it, and called out for "Some of you men of Co. F with guns to come up here."

His call was obeyed, and five hundred and six Confederates surrendered to this little company alone. In the company the casualties were as follows: Patrick Murray, killed; Eugene Mead, Watson P. Morgan and Fitz Green Halleck, wounded. Having thus uncovered the ford the sharpshooters were pushed forward some distance to allow the remainder of the left wing to cross and form on the south bank. Advancing about a mile from the river they took up a position from which they repulsed several feeble attacks during the day, and at dark were relieved.

For their gallantry and dash in this affair they received unstinted praise from their brigade commander, De Trobriand, they having been transferred back to his brigade some days previous. On the next day the troops advanced towards Brandy Station where the union of the two wings of the army was expected to take place. Considerable resistance was met with at several points during the day, and at one point the skirmishers of the third division, which was in advance, being unable to start the rebels, the corps commander sent back his aide for "the regiment that crossed the river the day before," but the brigade was some miles in rear of the point of obstruction, and Gen. De Trobriand, rightly believing that it would be unjust and cruel to require these men to march so far at the double quick after their severe service of the day before, sent the second regiment instead, who fully met the requirement and soon cleared the road for the head of the column.

On arriving at Brandy Station the vast open plain was found packed and crowded with troops, the entire Army of the Potomac being now concentrated here. The sharpshooters went into camp on the farm of the so called loyalist John Minor Botts, where they remained for the eighteen days following. In consideration of his supposed loyalty, every effort was made to protect the property of the owner of the plantation, but *rails* are a temptation that no soldier was ever known to withstand on a cold November night. Evil disposed troops of other organizations raided the fences every night, and the troops nearest at hand, the sharpshooters, were required to rebuild them every day; and in this manner they passed the time until the 26th of November, when the army broke camp and crossed the Rapidan at several points simultaneously.

This was the initial movement in what is known as the Mine Run Campaign. The Third Corps crossed at Jacobs Mills Ford, their destination being understood to be Robertson's tavern where they were to join the Second Corps in an attack on the Confederate line behind Mine Run at that point. But Gen. French, by a mistake of roads, and sundry other unfortunate errors of judgment, found himself far to the right of his assigned position, and while blindly groping about in the mazes of that wilderness country, ran the head of his column against Ewell's Corps and a brisk fight took place, which was called the battle of Locust Grove.

De Trobriand's brigade was near the rear of the column and was not therefore immediately engaged. The familiar sounds of cannon and musketry indicated to their practiced ears something more than a mere affair of skirmishers, and soon came an order to take up a more advanced position in support of the Third Division which was said to be heavily engaged. Upon arriving at the front the sharpshooters were deployed and ordered forward to a fence a little distance in advance of the main Union line, and to hold that position at all hazards. Moving rapidly forward they gained the position, and quickly converted the stout rail fence into a respectable breastwork from which they opened fire on the rebels in their front. Near them they found the Tenth Vermont, and thus once again stood shoulder to shoulder with the men of their native state. Five times during that afternoon did the enemy endeavour to drive the sharpshooters from this line, and as often were they repulsed, and each time with heavy loss. In one of these assaults the colours of a rebel regiment, advancing immediately against Co. F, fell to the ground four times, and just there four rebel colour bearers lay dead, stricken down by the fire of the Green Mountain riflemen.

The line of breastworks were held until the fighting ceased after dark, when the sharpshooters were relieved and retired from the immediate front and lay on their arms during the night. Co. F had lost in the battle of the day five good men; E. S. Hosmer was killed at the fence, while A. C. Cross, Eugene Payne, Sherod Brown and Corporal Jordan were wounded. Cross rejoined the company and served faithfully until the battle of the Wilderness in the following May where he was killed. Payne returned to duty and served his full term of enlistment and was honourably discharged on the 13th of September, 1864. Brown never fully recovered from the effects of his wound and was subsequently transferred to the Veteran Reserve Corps. Jordan also reported again for duty and served until the 31st of August, 1864, when he was honourably discharged on surgeon's certificate of disability. The regiment had lost thirty-six men killed and wounded during the day, while the corps had suffered a total loss of fifteen hundred, and had not yet reached its objective point. And this was the soldiers' Thanksgiving Day at Locust Grove. Far away in quiet northern homes, fathers and mothers were sitting lonely at the loaded tables thinking lovingly of their brave boys, who were even then lying stark and cold under the open sky, or suffering untold agonies from cruel wounds. But this was war, and war is no respecter of time or place, and so on this day of national thanksgiving and praise, hundreds of the best and bravest suffered and died that those who came after them might have cause for future thanksgiving.

> To the misjudging, war doth appear to be a worse calamity than slavery; because its miseries are collected together within a short space and time as may be easily, at one view, taken in and perceived. But the misfortunes of nations cursed by slavery, being distributed over many centuries and many places, are of greater weight and number.

Further severe fighting took place on the next day, but the sharpshooters were not engaged. On the twenty-ninth (the corps having changed its position on the previous day, taking up a new line further to the left), the sharpshooters were deployed as skirmishers and pushed forward to within sight of the strong works of the enemy on the further side of Mine Run where they were halted and directed to closely observe the movements of the rebels, but to do nothing calculated to provoke a conflict, the preparations for assault not being completed on the Union side. While laying here in a cold November rain storm they had ample opportunity to calculate the strength of the enemy's line and the chances of success.

It reminded them strongly of Fredericksburgh. The position was not dissimilar to that. Here was a swampy morass instead of a hard plain, but beyond was a height of land and, as at Fredericksburgh, it was crowned with earth works, while at the base of the elevation, plainly to be seen by the watchers, were the long yellow lines that told of rifle pits well manned by rebel soldiers. It looked like a desperate attempt, but early on the morning of the thirteenth, in obedience to orders, the sharpshooters advanced across the swamp through the partly frozen mud, in many places mid-leg deep, driving the rebel pickets into their works and pressing their way to within a few rods of the enemy's front, which position they held, being of themselves unable to go further without support, which was not forthcoming. This advance had the seeming character of a demonstration only, but the sharpshooters made the best of their opportunities, picking off a rebel now and then as the chance occurred. Night came on and no hint of relief came to the worn and weary men.

It was intensely cold and, of course, they had to endure it as best they could, since to light a fire within so short a distance of the watchful rebels would be to draw the fire of every gun within range. Neither could they get the relief which comes from exercise, for the first movement was the signal for a shot. So passed the long and dismal night; the men getting such comfort as they could from rubbing and chafing their benumbed and frost-bitten limbs.

Morning dawned, but yet no relief from their sufferings; and it seemed to the waiting men that they were deserted. At times firing could be heard on the right, but of other indications of the presence of their friends there were none. They remained in this state all day on the 1st of December, and at night, after thirty-six hours of this exposure, they were ordered back across the swamp. Many men were absolutely unable to leave their positions without aid, so stiff with cold and inaction were they; but all were finally removed. The army had retired from the front of the enemy and was far on its way to the river, leaving the Third

Corps to cover the withdrawal; the greater portion of this corps was also en route for its old camp, and the sharpshooters were thus the rear guard of the army.

The march was simply terrible. All night they struggled on, many men actually falling asleep as they marched and falling to the ground, to be roused by shakes and kicks administered by their more wakeful comrades. In spite of all, however, many men left the ranks and lay down in the fields and woods to sleep, preferring the chance of freezing to death, or of that other alternative only less fatal—being made prisoners—to further effort.

At day break the regiment arrived at the Rapidan at Culpepper Mine Ford, crossing on a pontoon bridge and going into bivouac on the north bank, where they could at least have fires to warm their half frozen bodies. Here they lay until noon, their numbers being augmented by the arrival of the stragglers, singly and in squads, until all were accounted for, though at day break there were not guns enough in some of the companies to stack arms with. At night, however, all were comfortably quartered in their old camp—a thankful lot of men. This was perhaps the most severe experience that Co. F had to undergo during its three years of service. On many occasions they had more severe fighting and had often to mourn the loss of tried and true comrades; but never before or after did the company, as a whole, have to undergo so much severe suffering as on this occasion.

The principal loss of the regiment in this campaign was by the death of Lieut.-Col. Trepp, who was shot through the head and instantly killed on the 30th of November. Col. Trepp had been with the regiment from the first, having joined as captain of Co. A. He was a Swiss by birth, and had received a military education in the army of his native land, and had seen much service in various European wars. He was a severe disciplinarian, even harsh; but was endeared to the men by long association in the field, and was sincerely lamented.

From this time until the 6th of February, 1864, the regiment lay in camp, inactive. On that day they were engaged in a reconnaissance to the Rapidan, but were not engaged.

On the 28th of March the gallant old Third Corps, reduced as it was by its losses at Chancellorsville, Gettysburg and Locust Grove to the proportions of a small division, passed out of existence, being consolidated with the Second Corps, and becoming the first and second brigades of the Third Division of that corps, Gen. Birney continuing in the command of the division, while the corps was commanded by Gen. Hancock, who had so far recovered from his wound received at Gettysburg as to be able to resume his place at the head of his troops. The sharpshooters were attached to the second brigade, commanded by Gen. Hays.

This change was viewed by the officers and men of the Third Corps with great regret. They were proud of their record, and justly so, but the necessities of the service were paramount, and no sentiment of loyalty to a corps flag could be allowed to interfere with it. In recognition of the distinguished services rendered by the old organization, however, the men were allowed to retain their corps badge; and they took their places in the ranks of Hancock's command resolved that the honour of the old Third should be maintained unsullied in the future, as it had been in the past.

CHAPTER 6

The Wilderness, Spotsylvania and Cold Harbor

On the 10th of March an order was received from President Lincoln assigning Gen. U. S. Grant to the command of all the armies of the United States, and during the last days of the same month Gen. Grant pitched his headquarters tent at Culpepper court house, and commenced a study of the situation in Virginia, where the real struggle of the war had been maintained for nearly three years, and where the strength of the Confederacy yet lay. The time, until the 3rd of May, was spent in active preparation for the opening of the spring campaign. Sick and disabled men were sent to the rear. All surplus baggage and stores were turned in, and the army, stripped for the fight, stood ready whenever the new commander should sound the advance; for although Gen. Meade was still commander of the Army of the Potomac, every man knew that Gen. Grant was there for the purpose of personally directing its movements. On the 3rd of May the sharpshooters broke camp and marched out on that campaign which was destined to be one continual battle for nearly a year to come, and at the end of which was to come the final triumph at Appomattox.

The organization of Co. F at this time was as follows:

Captain, C. D. Merriman; vice E. W. Hindes honourably discharged on surgeon's certificate of disability.

> First Lieutenant   H. E. Kinsman.
>
> First Sergeant   Lewis J. Allen.

Second Sergeant	Cassius Peck.
Third Sergeant	Paul M. Thompson.
Fourth Sergeant	L. D. Grover.
Fifth Sergeant	Edward F. Stevens.
First Corporal	Chas. M. Jordan.
Second Corporal	Edward Trask.
Third Corporal	M. Cunningham.
Fourth Corporal	Edward Lyman.
Fifth Corporal	D. W. French.
Sixth Corporal	Carlos E. Mead.
Seventh Corporal	Henry Mattocks.
Eighth Corporal	Chas. B. Mead.

With this organization and forty-three enlisted men, the company crossed the Rapidan at Ely's ford at nine o'clock a.m. on the 4th of May, 1864. Marching rapidly to the southeast; they bivouacked for the night near Chancellorsville on the identical ground on which they had fought exactly one year before under Hooker. The omen was not a happy one, but with high hopes of success under this new western general who had always beaten his enemies hitherto, they lay down prepared for whatever of good or ill the morrow might bring forth.

Reminders of the conflict of May, 1863, were thickly scattered about on the ground, and some men in the regiment found their hair covered knapsacks where they had thrown them off in the heat of the former battle, and which they had been forced to abandon. They found also the graves of some of their lost comrades, buried where they fell, while in many places human bones shone white and ghastly in the moonlight. It was the very ground over which the sharpshooters had driven the Stonewall brigade on the night of the 3rd of May of the preceding year. With the earliest streaking of the eastern sky on the morning of the fifth, the Second Corps, with the sharpshooters in the advance, was put in motion towards Shady Grove church, situated some four or five miles to the southward at the junction of two important roads, and where they were to form the extreme left of the army. Before the head of the column had reached that point heavy firing was heard on the right and rear, and the column was counter-marched and ordered to return to the junction of the Brock road with the Orange plank road, which the enemy were making desperate efforts to secure. It was indeed a matter of the utmost importance to maintain possession of the Brock road, since it was the very key to the whole battle ground. Running nearly north and south from the Orange turnpike, near the old Wilderness tavern, it intersects all the roads leading from the direction from which the enemy were approaching, and, as it is the only important, or even passable, road running in that direction, its possession by either army would enable that party to outflank the other almost at pleasure. Getty's division of the Sixth had been detached from that corps on the right some hours before, and ordered to hold this position at all hazards, and it was the sudden attack on this isolated command that had called the Second Corps back from its march towards Shady Grove church.

At about two o'clock p.m. Birney's division arrived at the threatened point and were at once deployed for action on the Brock road, and to the left, or south, of its intersection with the plank road. Here the men of Co. F. found themselves again shoulder to shoulder with their friends.

The old Vermont brigade formed part of Getty's division and were already deployed and sharply engaged; so that Co. F. found themselves in the immediate neighbourhood of the gallant Vermonters. Immediately upon the arrival of the head of the division upon the field, and pending the necessarily slower formation of the main line, the sharpshooters were pushed out towards the enemy and at once came under a heavy fire. It was their first fight under Hancock, and they felt that not only was their own well earned reputation to be sustained, but that the honour of the now dead and gone Third Corps was in a measure committed to their keeping. There, too, just on their right stood the men of the old brigade, proud of their own glorious record, and just a little inclined to rate their own courage and skill above that of any other troops in the army.

Under the stimulus of these conditions the sharpshooters as a regiment, and the men of Co. F in particular, fought with a dash and energy which surprised even their own officers who had learned long before that there was almost no task which the rank and file thought themselves unequal to. This contest of a skirmish line against lines of battle continued for nearly two hours; but at about four o'clock p.m., the whole of the Second Corps having arrived and being in position, a general advance was ordered, and now the fighting, which had been very severe before, became simply terrific. The ground was such that the artillery could not easily be brought into action. Only two guns could be brought up, which were placed on the plank road where they rendered excellent service. The musketry, however, was continuous and deadly along the whole line. The roar of battle was deafening, and struck upon the ear with a peculiar effect from the almost total absence of artillery, usually so noisy an accompaniment of modern battle. The men who noted this fact, however, were men accustomed to warfare, and who knew that the fire of infantry was much more deadly than that of artillery, and never before had they heard such continuous thunder or confronted such a storm of lead as on this occasion. The fierce struggle continued with unabated ferocity until the merciful night put an end to it. The Brock road was held, but it had been impossible to do more. The enemy were badly shattered, and at points the line had been broken; but the nature of the ground was such as to prevent an orderly and systematic pushing of such advantages as were, here and there, gained, and, except that the key point remained in the hands of the federals, it was a drawn battle.

The men lay on their arms during the night, in the position in which the cessation of the battle found them; and, as illustrative of the closeness of the contending lines, and the labyrinthine character of the ground, it may be stated that during the night many men from both armies while searching for water, or for their wounded friends, strayed within the opposing lines and were made prisoners. Among the above were Sergt. Paul

M. Thompson and J. H. Guthrie of Co. F. Besides these two men, Co. F had lost terribly in killed and wounded during the day. Corporal David French, W. J. Domag and E. E. Trask were killed on the field; A. C. M. Cross and Wm. Wilson were mortally wounded, while M. Cunningham, Spafford A. Wright, John C. Page, S. M. Butler and Wm. McKeever suffered severe and painful wounds—a total of twelve men lost out of the forty-three who answered to the roll call on that morning, and this in the first fight of the campaign.

But the survivors felt that they had well and nobly sustained the honour of their corps, and of their state. They were proud, also, to have received the commendation of distinguished officers of the old Vermont Brigade, and so, with mingled emotions of sorrow and gladness, they lay down on the bloody field. It will be remembered that the sharpshooters had been pushed out on the left of the plank road immediately upon their arrival and while the troops of the line were being formed on the Brock road. In this formation, Birney's division had been sent to the north or right of the plank road, and formed on Getty's right; so that during the subsequent battle the sharpshooters had been separated from their brigade, and had been fighting in an entirely independent manner, subject to no orders but those of their regimental and company officers. At daylight the men were rallied on the colours and moved to the north of the plank road in search of their proper command, which, after some search in the tangled forest, they found the shattered remains of. The brigade commander, Gen. Alexander Hays, and very many other gallant officers and men had fallen on the preceding day, and so heavy had been the losses that the entire brigade when deployed, hardly covered the front of an average regiment as they had stood when the army crossed the Rapidan.

Notwithstanding his severe losses of the day before, Gen. Grant (who, by the way, was understood to have expressed the opinion at some time that "The Army of the Potomac had never been fought up to its capacity") ordered another general assault along the whole line at five a.m. on the sixth.

Promptly at that hour the Second Corps advanced along the Orange plank road, the sharpshooters being now on the right of that thoroughfare with their own division. They were, as on the day before, in the front line, but on this occasion they were heavily supported from the start, Birney's and Mott's divisions being in the first line while Getty's division formed a second line, the whole supported by Carroll's and Owen's brigades of the Second Division of the same corps.

The attack was made with great vigour and impetuosity, and was for a time successful, the enemy being driven with great loss and disorder from two strong lines of works, one about four hundred yards behind the other, which they had materially strengthened during the night. Birney's left, in front of which was Co. F, advanced further than his right, driving the Confederates before them and completely disrupting their line at this point; in fact so far did they penetrate that they were in a position to take the rebel left in flank and rear, and at one time the sharpshooters, during a momentary lull on their own front, turned their attention to a Confederate battery which was actually in rear of their right, and which they had passed beyond in their charge. They were not destined to reap the fruits of this victory, however, for at this time Longstreet's command arrived on the field and commenced a furious attack on Birney's exposed left. Changing fronts to meet this new enemy, the sharpshooters, with the aid of their comrades of Birney's division, made a vigorous resistance to this counter attack. The momentum of their own charge was gone; they had now fought their way through nearly a mile of thickets and swamps and had, necessarily, lost their alignment and cohesion. The utmost they could now hope to do was to beat back the oncoming rebels and give the Union troops time to reform for another assault. It was a vain effort, for the fresh masses of rebel troops succeeded in forcing the advanced left back as far as the centre and right, which was at the same time, about seven o'clock a.m., struck by a strong force of Confederates. By desperate effort the line was held and a reorganization effected, and at about nine o'clock the offensive was resumed along the plank road. The force of this attack was seriously impaired by the supposed necessity of protecting the extreme left which was greatly exposed. For some time heavy firing had been heard in that direction, and ugly rumours of columns of infantry, too strong to be checked by the cavalry, were rife. Then, too, a considerable body of infantry was discovered actually approaching the left and rear from the direction of Spotsylvania. All this necessitated the detachment of considerable bodies of troops to guard that wing, which weakened the force of the main attack. The infantry force which had occasioned so much uneasiness proved to be a body of convalescents trying to rejoin the Union army, and the troops sent to oppose them were restored to the point of action. By this time, in the movement of the lines, the sharpshooters found themselves, with most of the division, again on the left of the plank road. The fighting now became as close and severe as that of the preceding day; so dense and dark was the thicket, that the lines were often close together before either could determine whether the other was friend or foe; regiments lost their brigades and brigades their divisions. Indeed, so confused was the melee that it is stated that one regiment, being surrounded and ordered to surrender, actually laid down their arms to another regiment of their own brigade.

Still, progress was made, and, on the whole, the federals, although losing heavily, were gaining substantial ground. After half an hour of this work the troops on the right of Birney's division having given way, Birney detached two of his own brigades to fill the gap, and at about eleven o'clock the resistance in front of Hancock's corps having nearly ceased, another halt was called to readjust the confused and irregular lines. Before this could be accomplished a new enemy appeared square on the left of Birney's division, which was doubled up by the suddenness and impetuosity of the attack, and the confusion became so great along the whole line that Gen. Hancock directed a withdrawal of the entire corps to the breastworks which had been constructed on the Brock road, and from which they had advanced on the day before. It began to look like the same old story—as though Chancellorsville was to be repeated—and as though the most cheerful bulletin Grant would have to send North would be the often repeated one, "The Army of the Potomac is again safe across the Rapidan."

But there, some way, seemed to be no actual movement looking in that direction—in fact, *Grant had ordered the bridges removed as soon as the last troops had crossed the river*, and for twenty-four hours there had been no possibility of re-crossing had any one been so minded. Lines of retreat seemed to have no place in the plans of the new general-in-chief.

The enemy followed the retiring Union troops closely, but once within the breastworks the Second Corps was soon rallied, and, reforming, lay down behind the rude entrenchments to await the signal for renewed action. The Confederates pushed their lines to within two or three hundred yards of the Brock road, but rested at that point until about four o'clock p.m., when they took the offensive in their turn and made a gallant assault on Hancock's command behind the breastworks. This attack was understood to be under the immediate direction of Gen. Lee, who was present and commanded in person.

The rebel line came gallantly forward to within a few yards of the road, when they halted and opened a fierce fire, which was returned by the Union troops from their shelter, coolly and with deadly effect.

Here the sharpshooters had the unusual good fortune to fight in a sheltered position instead of in the open field, as was usually their fate. During this affair the woods took fire and for a long time the troops fought literally surrounded by the flames. The wind was from such a direction as to bring the smoke from the blazing woods directly in the faces of the federal soldiers, while the heat and smoke combined made the position almost untenable, even had there been no other enemy to contend with. In many places the log breastworks themselves took fire and became a blazing mass which it was impossible to quench. Still the battle raged; at some points it was impossible to fire over the parapet, and the defenders were compelled to withdraw for a short distance. The rebels were prompt to take advantage of such breaks, and at one point pushed their advance up to and over the road, planting their battle flags on the Union works, but a brigade of Birney's division charged them with such vigour that their holding was of short duration and they were driven back in great confusion, leaving numbers of their dead and wounded inside the breastworks.

In this charge the sharpshooters were conspicuous. Advancing in line of battle and at the double quick, they forced the enemy from their front over and far beyond the road, pursuing them and making prisoners even beyond the lines which had been held by the rebels previous to their assault. Their regimental flag was the only one advanced beyond the line of works; other troops contenting themselves with simply repossessing the line of the road. In this charge Jacob Lacoy of Co. F. was killed, the only casualty in the company on that day. Following this repulse Grant, still aggressive, ordered another attack by Hancock, and the troops were formed for that purpose; but before the advance actually commenced the order was countermanded and the men of the Second Corps lay down for the night along the road which they had so gallantly defended. The morning of the third day of the battle opened with the greater portion of the army quietly resting on their arms; but for the sharpshooters there seemed no relief or respite. At day break they were deployed, again on the right of the plank road, and advancing over the scene of the fighting of the two previous days, now thickly covered with the dead of both armies, encountered the rebel skirmishers at a distance of about four hundred yards from the Union line. Ordered to halt here and observe the enemy, they passed the time until about noon in more or less active sharpshooting and skirmishing.

At twelve o'clock they were ordered to push the enemy back and develop if possible his main line. Supported by infantry they dashed forward and after sharp fighting drove the rebels back into their works, some half a mile away. Here they were brought to a halt and found themselves unable to advance further. Counter-attacks were made by the rebels which were for a time successfully resisted; but the regiment was at last so far outflanked that it became necessary to fall back to avoid the capture of the entire command. The rebels did not pursue vigorously; the fight was out of them, and with a few unimportant affairs on different portions of the line the day passed without battle. Neither party had won a victory. Grant had not destroyed Lee's army, neither had Lee driven Grant back across the river, as he had done so many other Union commanders, and the battle of the Wilderness was of no advantage to either party, save the fact that Grant had destroyed a certain number of Lee's soldiers who could not easily be replaced, while his own losses could be made good by fresh levy from the populous North. Whatever may have been Gen. Grant's idea of the "capacity" of the Army of the Potomac for fighting hitherto, or whether he believed it to have been now "fought up to its capacity," he was forced to acknowledge that the fighting of the past three days had been the severest he had ever seen. But his thoughts were not yet of retreat; he had seen enough of the Wilderness as a battle field, however, and on the evening of the seventh issued his orders for a concentration of his army on Spotsylvania.

Company F. had lost in the action of this day Edward Giddings and Joseph Hagan, killed, and Lieut. Kinsman, Dustin R. Bareau, Henry Mattocks and Edward Lyman, wounded. The wound received by Mattocks, although painful, was not such as to disable him, and he remained with the company only to lay down his life on the bloody field of Spotsylvania a week later. The total losses now footed up nineteen men since the morning of the 5th of May.

All night long columns were marching to the southward. It was evident that the army was to abandon this battle field, but it seemed strange that the customs and traditions of three years should be thus ruthlessly set aside by this new man, and that he should have turned his face again southward, when by all precedent he should have gone north. The men, however, began to surmise the true state of affairs, and when during the night

Grant and Meade, with their respective staffs, passed down the Brock road headed still south, the men took in the full significance of the event, and, tired and worn as they were, they sprang to their feet with cheers that must have told Grant that here were men fully as earnest, and fully as persistent as himself in their determination to "fight it out on that line." The stench from the decomposing bodies of the thousands of dead lying unburied filled the air and was horrible beyond description, and the sharpshooters were not sorry when at nine a.m., on the morning of May 8th, they were relieved from their duties on the picket line and, forming on the Brock road, took up their line of march toward Spotsylvania. They were the last of the infantry of the whole army; a small body of cavalry only being between them and the rebels who might well be expected to pursue.

The cavalry soon found themselves unable to check the pursuers, and Co. F, now the rear guard of the army, was faced about and deployed to resist the too close pursuit. In this order, and constantly engaged with the rebel cavalry following them, they retired fighting, until at Todd's tavern they found the rest of the division. During the day Wm. Wells was wounded and taken prisoner, the only casualty in the company during the day. Wells met the same sad fate which befell so many thousands of unfortunate prisoners, and died at Florence, S. C., during the month of September following.

Immediately upon their arrival a portion of the regiment, including Co. F, was placed on the picket line to the west of the tavern, their line extending across the Catharpin road. Here they met the advance of Early's rebel corps, and some skirmishing took place; but the rebels were easily checked, and no severe fighting took place. Early on the morning of the ninth a strong force of the enemy's cavalry appeared in their front and made a vigorous effort to force a passage. They were strongly resisted and at last forced to retire before the well aimed rifles of the Vermonters. Following rapidly, the sharpshooters pushed them to and beyond the Po River, along the banks of which they halted.

During this affair a rebel captain of cavalry was wounded and captured. Capt. Merriman, whose sword had been shot from his side during the action of the preceding day, thinking that a fair exchange was no robbery, appropriated the captured rebel's sabre, and thenceforth it was wielded in behalf of instead of against the Union. In the afternoon of this day the sharpshooters were recalled from their somewhat exposed position, more than two miles from any support, and resumed the march towards Spotsylvania, skirmishing with the rebels as they retired, until they reached the high around overlooking the valley of the Po, where they found the rest of the corps making preparations to force the passage of the river.

The Union artillery was noisily at work, while rather faint response came from the enemy on the opposite side. A rebel signal station was discovered some fifteen hundred yards away, from which the movements of our troops could be plainly observed, and from which Gen. Hancock desired to drive the observers. A battery opened fire on them, but the distance was too great for canister, and the saucy rebels only laughed at shell. The men of Co. F., who were in plain view of both parties, watched this effort with great interest for half an hour, when they concluded to take a hand in the affair themselves. Long practice had made them proficient in judging of distances, and up to a thousand yards they were rarely mistaken—this, however, was evidently a greater distance than the rifles were sighted for. They therefore cut and fitted sticks to increase the elevation of their sights and a few selected men were directed to open fire, while a staff officer with his field glass watched the result.

It was apparent from the way the men in the distant tree top looked *down* when the Sharpes bullets began to whistle near them that the men were shooting under still, so more and longer sticks were fitted to still further elevate the sights; now the rebels began to look *upward*, and the inference was at once drawn that the bullets were passing over them. Another adjustment of the sticks, and the rebels began to dodge, first to one side and then to another, and it was announced that the range was found. Screened as they were by the foliage of the tree in which they were perched, it was not possible to see the persons of the men with the naked eye; their position could only be determined by the tell-tale flags; but when all the rifles had been properly sighted and the whole twenty-three opened, the surprised rebels evacuated that signal station with great alacrity. Gen. Hancock had been a close and greatly interested observer of this episode, and paid the men handsome compliments for their ingenuity and skill. The same night the division commander, Gen. Birney, ordered that thereafter the sharpshooters should report directly to his headquarters and also receive their orders from the same source. They were thus detached from their brigade.

At six o'clock p.m. the line advanced, and, after some slight resistance, effected the passage of the river. Pushing forward the sharpshooters soon found themselves again on the banks of the same river, which here changes its course to the south so as to again cross the road along which the corps was advancing. It was now well into the night, and as the men found the river too deep to ford; the column was halted and spent the night in this position. The second corps, which had held the entire left of the Union line ever since the crossing of the Rapidan a week before, by these manoeuvres found itself now on the extreme right of the army, and its position was a serious menace to Lee's left flank.

Indeed Barlow's division, as it lay that night, was actually in rear of the rebel left. Lee was quick to perceive the seriousness of the situation, and during the night he placed a formidable force in Hancock's front, and by the morning of the eleventh the corps found a strong line of works, well manned, to oppose their further progress. Reconnaissances were made, and a crossing effected at a point lower down, but the position was deemed too strong to attack, and the troops who had crossed were retired, soon after which the entire command was withdrawn to the northern bank of the Po.

Birney's division was first over, and thus escaped the severe fighting which befell the other portions of the command in the movement. During all this time the battle had been raging furiously on the centre and left of the Union army; repeated desperate assaults had been made at various points, and everywhere the enemy were found in great force behind strong works. The different assaults had been bloodily repulsed and the losses of men had been terrible. Still there was no sign of a retrograde movement. Grant seemed to have an idea that the true course of the Army of the Potomac lay to the southward instead of to the north. A repulse—such as would have been to the former commanders of that army a defeat—only spurred him to renewed effort, and it was in the evening of this day that he sent to President Lincoln the celebrated dispatch which so electrified the people of the North and made it clear to them that thenceforth there were to be taken no steps backward. "I propose to fight it out on this line if it takes all summer."

The operations of the past two days had convinced Generals Grant and Meade that a salient near the centre of Lee's entrenched line was his weakest point, and during the afternoon and night of the eleventh the troops selected were brought up and formed for the assault. The point at which the attack was aimed was the one which has since come to be called the Death Angle at Spotsylvania; and well was it so called. Hancock's command was withdrawn from the extreme right and placed on the left of the Sixth Corps in such a position that their advance would bring them, not opposite the exact angle, but on the rebel right of that point. Birney's division had the right formed in two lines of battle, with Mott's division in one line in support. The sharpshooters were deployed on the right of Birney's front line so as to connect the right of the Second Corps with the left of the troops next on the right. The night was made doubly dark by a thick fog which shut out all objects from sight at a distance of even a few yards, and in groping along to find their designated position, the men found themselves far in advance of the proper point and close up to the rebel line. As soon as their presence was discovered the enemy opened a brisk fire upon them, but believing their position to be at least as advantageous as the one they had left behind, the men lay quietly down without replying to the enemy and waited the signal of attack. They were now exactly opposite the Death Angle and only a few yards from the abatis. At half past four a.m. the signal was given, and the troops of the main line, rising to their feet, moved forward silently to the attack.

The sharpshooters, far in the advance, lay quietly until the charging lines were abreast of them when they too sprang up and dashed straight at the enemy's works. The lines were now in entirely open ground, sloping upward toward the enemy, and fully exposed to the fire which came thick and deadly from every gun that could be brought to bear. Men fell rapidly, but nothing could stay the magnificent rush of the veterans of the Second Corps, and with ringing cheers they crowned the works with their standards and fairly drove the rebels out by the sheer weight and vigour of their charge. Not all, however—for nearly four thousand Confederates, including two general officers, surrendered themselves as prisoners. Some thirty colours and twenty guns were also captured.

The sharpshooters were active in the assault and also in the short pursuit, which was brought to a sudden check, however, by the sight of a second line of works extending across the base of the triangle made by the salient. The Union troops were now a confused mass of rushing men. They had lost their brigade, regimental and even their company organization, as not infrequently happens in such assaults, and the enemy, advancing from behind their second line, compelled the triumphant but disordered federals to retire to the captured works where they were rallied. Quickly reversing the order of things, they, in their turn, became the defenders where they had so lately been the attacking party. Forming on the exterior slope, they fought the rebels stubbornly. It was as apparent to Lee as it had been to Grant and Meade, that this was the vital point, and now both parties bent their utmost energies—the one to hold what they had gained, and the other to repossess themselves of what they had lost.

Both lines were heavily reinforced and the fighting assumed the most sanguinary character of any that had been seen during the whole of the bloody three years of the war.

With desperate valour the Confederates rushed again and again against the Union lines to be met with a fierce fire at such short ranges, and into such dense masses, that every shot told. In some places they gained the crest of the breastworks and savage hand to hand encounters took place, but it was in vain; not all the valour of the boasted chivalry of the South could pass that line. Those who gained the works could not stay and live, and to retreat was as bad. Many gave themselves up as prisoners, while others, taking shelter on the other side of the works, kept up the fight by holding their muskets high above their heads and thus firing at random among the Union troops on the reverse side.

All day long this terrible combat continued. The dead on each side lay in heaps—literally piled the one on the other, until in many places the ground was covered three and four deep. The very trees were cut off by musket balls and fell to the ground. There is in the War Department at Washington, to this day, the stump of a tree more than eighteen inches in diameter which was cut down by this awful fire. Darkness brought with it an abatement, but not a cessation of the struggle; for until three o'clock in the morning of the thirteenth the strife continued. At that hour the enemy definitely abandoned the attempt to recapture the angle and retired to an interior line. Twice during the day had Co. F exhausted the ammunition in its boxes, and it was replenished by a supply brought to them as they lay by the stretcher bearers, and once the regiment was retired for a fresh supply, upon receipt of which they returned to the fighting.

In this carnival of blood—this harvest home of death—Co. F again suffered the loss of brave men. Henry Mattocks, Thomas Brown and John Bowen were killed, and Amos A. Smith and J. E. Chase were wounded. Only eighteen men were now left out of the forty-three who entered the campaign; twenty-five had fallen on the field.

A great sovereign once addressed his general thus: "I send you against the enemy with sixty thousand men."

"But," protested the general, "there are only fifty thousand."

"Ah!" said the *Emperor*, "but I count *you* as ten thousand!"

So each man of the gallant few who were left of what had been Co. F agreed to call his comrade equal to two men, and so they counted themselves yet a strong company.

The night of the twelfth was spent on the line which had been won and held at such a fearful cost of life. At twelve o'clock on the thirteenth the regiment, now but a handful of men, were moved by the right flank some three or four hundred yards, and ordered to establish a picket line in front of this new position. This was successfully accomplished with but little opposition and no loss to Co. F. That evening they were relieved and returned to division headquarters, where they bivouacked for the night. The three succeeding days were spent in the same manner; out before daylight, establishing new picket lines, sharpshooting as occasion offered, and spending the night near headquarters; but no important affair occurred, and no casualties were reported.

The seventeenth was spent quietly in camp—the first day of uninterrupted repose the men had enjoyed since crossing the Rapidan two weeks before. During that eventful period there had not been one single day, and hardly an hour, that the men of Co. F had not been under fire. It was a short time to look back upon, but what a terrible experience had been crowded into it! The company which is the subject of this history had lost more than half of its numbers, while in the Army of the Potomac the losses had been appalling—no less than four thousand five hundred and thirty-two men had been killed on the field, and the wounded numbered eighteen thousand nine hundred and forty-five (a total of twenty-two thousand four hundred and seventy-seven men) while of the missing there were four thousand eight hundred and seventy-two, making a total of twenty-seven thousand three hundred and forty-nine lost from the effective strength of the army since May 4th. Some idea of the extent of the losses may be obtained by the casual reader by a comparison, thus: If the entire population of any of the great and populous counties of Bennington, Orange or Orleans, as shown by the census of 1880, were suddenly blotted out, the loss would not equal the total of killed and wounded during the twelve days between the 4th and 17th of May, while the entire population of Grand Isle county is not as great as the number of the killed alone; and the total loss in killed, wounded and missing is greater than the population of any county in the State of Vermont except Chittenden, Franklin, Rutland and Windsor. And yet there was no sign of retreat. On the contrary, on every side were evidences of preparation for renewed battle, and during these days of comparative quiet attempts were made at various points to penetrate the rebel line, some of these assaults rising of themselves almost to the dignity of battles, but so insignificant were they as compared with what had gone before that they hardly attracted the attention, even, of any but the men immediately engaged.

On the nineteenth Gen. Grant ordered another movement of the army, again by the left, and again in the direction of Richmond. No unusual incident occurred to mark the progress of the sharpshooters until the twenty-first, when the regiment, by a sudden dash, occupied the little village of Bowling Green, where the retreating enemy had confined in the jail all the negroes whom they had swept along with them, and whom they intended to remove to a point further south where they would be removed from the temptation to desert their kind masters and join the Union forces. The advance was too sudden for them, however, and some hundreds of negro slaves were released from their captivity by the willing riflemen.

Two miles beyond Bowling Green the skirmishers met a considerable force of rebel cavalry, and a sharp skirmish took place. Two regiments of new troops came into action on the right, but being dispersed and routed retired to be seen no more, and the sharpshooters fell heirs to their knapsacks which they had laid off on going into action. The departed regiments had evidently had a recent issue of clothing, and their successors were thankful for the opportunity of renewing their own somewhat dilapidated wardrobes. They were further gratified about this time by the arrival of four convalescents, which swelled the number to twenty-two for duty. The twenty-second was a red letter day for the men who had been confined to such rations as they could carry on their persons. On this day they were ordered on a reconnaissance which took them into a section of country not frequently visited by either army

Halting at the County Poor House, they proceeded to gratify a soldier's natural curiosity to see what might be found on the premises to eke out their unsatisfactory rations, and, to their great delight, found chickens, mutton, milk and eggs in profusion, upon which they regaled themselves to their hearts' content. If these, thought the delighted men, are Virginia poor house rations, the poor of Virginia are greatly to be envied. Proceeding on the twenty-third towards Hanover Junction, they found their way once again blocked by the rebel army in a strong position behind the North Anna River and prepared again to receive battle on a fortified line of their own choosing. This was a disappointment, for the soldiers had become tired of such work and ardently desired to get at the rebels in an open field; but Grant, patient and persistent as ever, at once set about finding a means whereby he might beat them even here, if such a thing was possible.

The line of march had brought the Second Corps to the extreme left of the army, and it struck the river at the point at which the telegraph road crosses it at the county bridge. Here the enemy had constructed, on the north side of the river, a strong work for the defence of the bridge head; while on the southern bank, completely commanding the approaches to the river, was another, and a still stronger line of fortifications. The land in front of the nearer of the two was a bare and open plain, several hundred yards in width, which must be passed over by troops advancing to the attack, and every foot of which was exposed to the fire of the enemy on either bank.

To Birney's division was assigned the task of assaulting this position, and at five o'clock p.m., on the twenty-third, the division moved out in the discharge of its duty, Pierce's and Egan's brigades in the front line, while the Third brigade formed a second, and supporting line. The sharpshooters were deployed as skirmishers and led the way. The works were won without serious loss, and the sharpshooters passed the night near the river, charged with the duty of protecting the bridge for the passage of the troops on the next day, Gen. Hancock not deeming it advisable to attempt the crossing at that late hour of the evening.

Attempts were made during the night by the rebels to destroy the bridge, but it was safely preserved, although the railway bridge below was destroyed, and on the morning of the twenty-fourth, the troops commenced crossing covered by the fire of the sharpshooters, who lined the north bank, and the Union artillery posted on the higher ground in the rear. The regiment followed the last of the troops, and were pushed forward beyond the Fox house, a large, though dilapidated Virginia mansion, where they met the rebel skirmishers.

Sharp firing at long range continued for some hours until the ammunition in the boxes became exhausted, when the regiment was relieved and fell back to the Fox house, where breastworks were thrown up and where they remained during the rest of that day and the next, exposed to desultory artillery fire, but suffering no considerable loss. The next day the quartermaster, Lieut. Geo. A. Marden, arrived with the regimental wagons, and with such stores, clothing, and so forth, as the small train could bring.

As it was the first sight the regiment had had of its baggage for twenty-two days, the arrival was the signal for great rejoicing among the men, especially as the good quartermaster brought a mail, and the heart of many a brave soldier was made glad by the receipt of warm and tender words from the loved ones far away among the peaceful valleys of the state he loved so well.

The morning of the twenty-sixth brought sharp fighting for the troops on the right and left, but in Birney's front all was quiet, and the tired sharpshooters lay still until dark, when they were ordered to relieve a portion of the pickets of the Ninth Corps on their right. The night was very dark, and it was with difficulty that they found their designated position; but it was finally gained and found occupied by the Seventeenth Vermont, among whom the men of Co. F found many friends.

During the night the army was withdrawn to the north bank of the river, and on the morning of the twenty-seventh the sharpshooters were also withdrawn, and operations on the North Anna ceased. Grant had found the position too strong to warrant another attempt like those of the Wilderness and Spotsylvania, and had determined on another movement to the left. All day, and until two o'clock the next morning, the troops toiled on, passing on the way the scene of a severe cavalry fight a few days previous, the marks of which were plainly visible to the eye as well as apparent to the nose, since the stench from the decaying bodies of horses and men was almost unbearable. After a few hours of needed rest the march was resumed at daylight, still to the south, and at four o'clock they crossed the Pamunkey at Hanover town. They were now approaching familiar ground. Only two or three miles away was the old battle field of Hanover court house, while but little further to the south lay Mechanicsville and Gaines Hill, where they had fought under McClellan two years before. Halting in a field near the river they rested until near noon of the following day.

During the forenoon of this day an inspection was had, from which it was inferred by some that it was Sunday, although there was no other visible sign of its being in any sense a day of rest. In the afternoon a reconnaissance in force was ordered to determine, if possible, the whereabouts of the rebels. Some skirmishing took place, but no important body of the enemy was found until the advance reached the point at which the Richmond road crosses the Totopotomy, where the enemy were found strongly posted with their front well covered by entrenchments and abatis, prepared to resist a further advance. A brisk skirmish took place, and the rebels were forced into their works.

The whole corps was now ordered up and took position as

close to the rebel line as it was possible to do without bringing on a general engagement, for which the federal commanders were not ready. In this position they lay, exchanging occasional shots with the rebel sharpshooters, but with little or no serious fighting, until the evening of June 1st, when the corps was ordered again to the left, and by a forced march reached Cold Harbor early in the forenoon of the second.

At two o'clock a.m. on the 30th of May Capt. Merriman had been ordered to take a detail of twenty-five men from the regiment and establish a picket line at a point not before fully covered. In the darkness he passed the proper position and went forward until he reached the rebel picket line, which, after challenging and receiving an evasive answer, opened fire on him. By careful management, however, he was able to extricate his little force, and eventually found and occupied his designated position. This was an unfortunate locality for Capt. Merriman, for when the corps moved on Cold Harbor, he, by some blunder, failed to receive his orders and was thus left behind.

Finding himself abandoned, and surmising the reason, he took the responsibility of leaving his post; and as it was clearly the proper thing to do under the circumstances, he escaped without censure. Severe fighting had already taken place between the Sixth and Eighteenth Corps and the rebels, for the possession of this important position, and *Old* Cold Harbor had been secured and held for the Union army. This little hamlet is situated at the junction of the main road from White House to Richmond, and the road leading south from Hanover town, which, a mile south of Old Cold Harbor intersects the road leading south-easterly from Mechanicsville, which road in its turn connects with the Williamsburgh road near Dispatch Station, on the Richmond & York River railway. The control of the road from White House was indispensable to the Union army, as it was the only short line to the new base of supply on the Pamunkey.

A mile to the westward of Old Cold Harbor this road intersects the Mechanicsville road at a place called *New* Cold Harbor, the possession of which would have been more desirable, since it would have given to the Union commander all the advantages of the roads heretofore mentioned and, also, the possession and control of the highway from Mechanicsville to Dispatch Station, which gave to the party holding it the same advantage which the Brock road had afforded to the Union troops in the Wilderness; that is, the opportunity to move troops rapidly over a good road, and by short lines, from right to left, or *vice versa*. This point was, however, held by the confederates in great force, and was defended by formidable works.

The heavy fighting of the day before had been for its possession, and the federals had not only gained no ground, but the troops engaged had suffered a disastrous repulse with severe loss, no less than two thousand men having fallen in the assault.

The morning of the 2nd of June brought to the anxious eyes of the federals the same familiar old view. In every direction across their front were seen the brownish red furrows which told of rifle pits, which at every commanding point in the rebel line rose stronger and higher works, above which peered the dark muzzles of hostile artillery.

It was evident that one of two things would ensue. Either a sanguinary battle, like those of the Wilderness and Spotsylvania, where the rebels, strongly entrenched, had all the advantages on their side must be fought, or Grant must try another move by the left and seek a more favourable battle ground. But that meant a move to the James River; since between the White House and the James there could be no new base of supply. Furthermore, the ground further to the south and nearer the James, was known to be fully as difficult as that on which the army now stood and was, presumably, as well fortified. And even if it was not fortified, the further Grant moved in that direction the stronger grew Lee's army, since the troops in and about Richmond, reinforced by a very large portion of those who had so recently made, and still kept, Butler and his thirty thousand men close prisoners at Bermuda Hundred, could be safely spared for more active operations in the field against this more dangerous enemy.

Moreover Grant had said "I propose to fight it out on this line," and it was now nineteen days since the fight for the angle at Spotsylvania, and the Army of the Potomac had hardly lost that number of hundreds of men in the operations on the North Anna and the Totopotomy. It was time to fight another great battle, lest the army should forget that it was now to be "fought up to its capacity," and so the battle of Cold Harbor was ordained.

The position of the Second Corps was now, as at the Wilderness, on the extreme left of the army; on their left were no forces, except the cavalry which watched the roads as far to the south as the Chicahominy. It was well remembered ground: two years before the sharpshooters, then part of the Fifth Corps, had, with that organization, fought the great battle of Gaines Hill, on this identical ground, but how changed was the situation.

They had now the same enemy before them, but the positions were completely reversed. Then, they were fighting a defensive battle for the safety of the army. Then, the enemy came far out from their entrenchments and sought battle in the open field. Now, it was the federals who were the aggressive party, and the rebels could by no means be tempted from the shelter of their strong works. Now, the enemy occupied nearly the same lines held by the federals on the former occasion, while the federals attacked from nearly the same positions, and over the same ground, formerly occupied by the rebels.

Then, however, the federals had fought without shelter; now, the rebels were strongly entrenched. Indeed, an unparalleled experience in warfare had taught both parties the necessity of preparation of this kind to resist attack, or to cover reverses. There was, however, a greater change in the moral than in the physical situation. Then, the rebels had been haughty, arrogant and aggressive; now, they were cautious and timid. Brought squarely to the test of battle they were, individually, as brave as of yore, but the spirit of confidence had gone out of them. They had learned at last that "one southern gentleman" was not "the equal of three northern mudsills."

The handwriting on the wall was beginning to appear plainly

to them, and while they still fought bravely and well—while they were still able to deal damaging blows, and to inflict terrible punishment—they never afterwards fought with the dash and fire which they had shown at Gaines Hill, at Malvern, at the Second Bull Run, at Chancellorsville, or at Gettysburg. The noontide of the Confederacy had passed, and they knew then that henceforth they were marching towards the darkness of the certain night.

The 2nd of June was spent by both parties in strengthening positions and other preparations. Constant firing, it is true, was going on all along the line, but no conflict of importance took place on this day. Co. F was thus engaged, but no important event occurred on their front. On the third, however, at half past four a.m. the corps moved forward to the assault. Barlow's and Gibbon's divisions formed the front line, while Birney's was in the second.

The early morning fogs still hung low and rendered it impossible for the advancing troops to see what was before them; thus many parts of the line became broken by obstacles which might have been, in part, avoided had it been possible to discover them in time, and the column arrived at the point of charging distance somewhat disorganized. Still the vigour of the attack was such that the rebels could not long resist it; they were driven out of a sunken roadway in front of their main line, into and over their entrenchments, and at this point the success of the assault was complete. Several hundred prisoners and three guns were captured, the guns being at once turned upon their former owners.

The supporting column, however, failed, as is so often the case, to come up at the proper time and the enemy, being strongly reinforced, advanced against the victorious men of the Second Corps, and after a desperate struggle, reminding the participants of the fight at Spotsylvania, forced them back and reoccupied the captured works. In this affair Co. F, being with Birney's division in the second line, was not actively engaged, nevertheless in the charge they lost two or three men whose names are not now remembered, slightly, and Alvin Babcock, mortally wounded. Babcock was one of the recruits who joined the company on the day after the battle of Antietam, nearly two years before, and had been a faithful and good soldier. He died on the first of July following from the effects of his wound.

The corps retired in good order to their own works. A partial attack by the rebels on their position was easily repulsed, and the rest of the day was passed in comparative quiet. The picket line, in full view of the rebel works and only about one hundred yards distant, was held by a regiment for whose marksmanship the rebels seemed to have a supreme contempt, since they exposed themselves freely, using the while the most opprobrious epithets.

The fire of their sharpshooters was constant and close, and a source of great annoyance to all within range. Co. F lay some distance in the rear of the pickets and somewhat exposed to the stray bullets which passed over the front line. They became somewhat restive under this unusual state of affairs; but receiving no order to move up to take part in the conflict, and having no liberty to shift their position, Capt. Merriman and Sergt. Peck determined to see what could be done by independent effort to relieve the situation. Taking rifles and a good supply of ammunition they made their way to the front and, taking up an advantageous position, commenced operations.

The first shot brought down a daring rebel who was conspicuously and deliberately reloading his gun in full view of a hundred Union soldiers. This single shot and its result seemed to convey to the minds of the rebels that a new element had entered into the question, and for a few moments they were less active. Soon regaining their courage, however, and apparently setting it down as the result of some untoward accident, they resumed their exposure of persons and their annoying fire. It did not long continue, however, for wherever a man appeared within range he got such a close hint of danger, if indeed he escaped without damage, that the sharpshooting along that front ceased. Further to the right was a place where the breastwork behind which the rebel infantry was posted did not quite connect with a heavy earthwork which formed part of the rebel line, and which was occupied by artillery. Across this open space men were seen passing freely and openly, apparently officers or orderlies passing along the line in the discharge of their duties.

To this point the two sharpshooters now directed their at attention. Dodging from tree to tree, now crawling along behind some little elevation of land, and now running at full speed across some exposed portion of the ground, they reached a place from which they could command the passage, and very soon the rebels found it safer and more convenient to take some other route. Service of this independent nature had a peculiar fascination for these men. In fact, sharpshooting is the squirrel hunting of war; it is wonderful to see how self-forgetful the marksman grows—to see with what sportsmanlike eyes he seeks out the grander game, and with what coolness and accuracy he brings it down. At the moment he grows utterly indifferent to human life or human suffering, and seems intent only on cruelty and destruction; to make a good shot and hit his man, brings for the time being a feeling of intense satisfaction.

Few, however, care to recall afterwards the look of the dying enemy, and there are none who would not risk as much to aid the wounded victim of their skill as they did to inflict the wound. War is brutalizing, but the heat of the actual conflict passed, soldiers are humane and merciful, even to their foes. The assault of the Second Corps had not been an isolated attempt to force the rebel line at one point only. On their immediate right the Sixth and Eighteenth Corps had also advanced, and had met with severe loss; while far away to the north, even to and beyond the Totopotomy, miles away, Burnside and Warren had been engaged in more or less serious battle. At no point, however, except in front of the Second Corps had the enemy's line been entered, and this lodgement, as has been seen, was of brief duration. Advanced positions had been held, however, and in many places a distance no greater than fifty to one hundred yards now separated the opposing lines. Barlow's division, magnificent fighters, when forced out of the captured rebel works, had taken advantage of a slight crest of ground not fifty yards from the rebel line, and with the aid of their bayonets, tin cups, etc., had thrown up a slight cover, from which they stubbornly refused to move; and to this far advanced line Companies F and G were ordered during the night of the third to keep down, so far as they were able, the rebel fire when the morning light should enable them to see the enemy. They spent the fourth in this position, constantly exposed and constantly engaged, suffering the loss of one man, Joseph Bickford, killed. The shooting on the part of the rebels was unusually close and accurate, and was a source of great discomfort to one, at least, of the men of Co. F. Curtiss Kimberly, known best by his friends as "Muddy," had such a breadth of shoulders that the small stump behind which he lay for shelter was insufficient to cover both sides at once. Three times in as many minutes the stump was struck by rebel bullets, and "Muddy" gravely expressed the opinion that there was "a mighty good shot over there somewhere," at the same time uttering an earnest hope that "he might not miss that stump."

During the night of the fourth they were moved to the left, and at daylight found themselves face to face with the rebel pickets near Barker's Mill. This was indeed "Tenting on the old camp ground," since this point had been the extreme right of the Union line at the battle of Gaines Hill, June 27, 1862.

They lay in this position until the twelfth, engaged every day, to a greater or less extent, in skirmishing and sharpshooting until the eleventh, when an agreement was made between the pickets that hostilities should cease in that part of the line, and the day was spent in conversation, games, etc., with the rebels. They were ravenous for coffee, but had plenty of tobacco. The federals were "long" of coffee but "short" of tobacco, and many a quiet exchange of such merchandise was made in the most friendly way between men who for days had been, and for days to come would be, seeking each others lives. It was a curious scene and well illustrated one phase of war.

On the twelfth, the truce being over, hostilities were resumed and the men who had so lately fraternized together were again seeking opportunity to destroy each other. On this day Almon D. Griffin, who had been wounded at Chancellorsville, was again a victim to bullets. He recovered, however, and rejoined his company to serve until the expiration of his term of service, when he was discharged. Grant was now minded to try another movement by the left, this time transporting his entire army to the south bank of the James, and on the thirteenth the sharp-shooters crossed the Chicahominy at Long Bridge, and leaving the old battle field of Charles City cross-roads and Malvern Hill to the right, struck the James River the same night at Wilcox's landing some two miles below Harrison's, where McClellan's army had lain so long after his unfortunate campaign in 1862. This was the first opportunity for a bath which had been offered since the campaign opened, and soon the water was alive with the dirty and tired men, their hands and faces of bronze contrasting strangely with the Saxon fairness of their sinewy bodies, as they laughingly dashed the water at each other, playing even as they did when they were school boys in Vermont. It was a luxury which none but those who have been similarly situated can appreciate.

CHAPTER 7

Siege of Petersburg, Muster Out

Early on the morning of the fourteenth the regiment crossed the James by means of a steam ferry boat and spent the day near the south bank. There was trouble somewhere in the quartermaster's department, and no rations could be procured on that day. On the next day orders were issued for an immediate advance; still no rations, and the hungry men started out on the hot and dusty march of some twenty miles breakfastless and with empty haversacks. But a hungry soldier is greatly given to reconnaissances on private account, he has an interrogation point in each eye as well as one in his empty stomach.

Every hill and ravine is explored, the productions of the country, animal and vegetable, are inventoried, and poor indeed must be the section that fails to yield something to the hungry searcher. Chickens, most carefully concealed in the darkest cellars by the anxious owners, are unearthed by these patient seekers, pigs and cows driven far away to the most sequestered valleys are brought to light; bacon and hams turn up in the most unexpected places, and on the whole, the soldier on a march fares not badly when left to his own devices for a day or so. Thus our sharpshooters managed to sustain life, and at dark went into bivouac in front of the rebel defences of Petersburg.

The Eighteenth Corps, under Gen. Smith, had preceded the Second, and had had heavy fighting on the afternoon of this day; they had captured and now held important works in the line of rebel defences. Darkness and an inadequate force had prevented them from following up their advantages, and thus the first of the series of terrible battles about Petersburg had ended.

At daylight on the morning of the sixteenth the Union artillery opened a brisk cannonade on the now reinforced enemy. During the forenoon the sharpshooters lay quietly behind the crest of a slight elevation in support of a battery thus engaged. At about noon they were deployed and advanced against the rebel pickets with orders to drive them into their main line and also to remove certain fences and other obstructions so as to leave the way clear for an assault by the entire corps at a later hour. The advance was spirited, and after a determined resistance the rebels were driven from their advanced rifle pits, the skirmishers following them closely, while the reserve companies levelled the fence in the rear.

At six o'clock p.m. the Second Corps, supported by two brigades of the Eighteenth on the right, and two of the Ninth on the left, advanced to the attack, and after severe fighting, in which the corps suffered a heavy loss in officers and men, they succeeded in capturing three redans in the rebel line of works, together with the connecting breastworks, and in driving the enemy back along their whole front.

Darkness put an end to the advance, but several times during the night the rebels attempted to regain their lost works, and were each time repulsed with loss. In this charge Caspar B. Kent of Co. F was killed on the field. Co. F moved during the night to a position further to the left, and farther to the front than any point reached by the Union troops during the day, and were made happy by an issue of rations, the first they had received since leaving the lines of Cold Harbor. A fresh supply of ammunition was also received by them, of which they stood in great need, they having very nearly exhausted the supply with which they went into the fight. The rebels in their front were active during the night and a good deal of random firing took place, but of course with little result so far as execution went.

Morning, however, showed a new line of rifle pits thrown up during the night, not over fifty yards in front of the sharpshooters who had by no means spent the night in sleep themselves, but in making such preparations for defence as they could with such poor tools as bayonets, tin plates and cups. They had been sufficient, however, and daylight found them fairly well covered from the fire of the enemy's infantry, and with a zigzag, or covered way, by means of which a careful man could pass to the rear with comparatively little danger. Co. F held this advanced line alone, and the day which dawned on them lying in this position was destined to be one of the most active and arduous, and the one to be best remembered by the men present, of any during their entire term of service.

No sooner did the light appear than sharpshooting began on both sides, and was steadily kept up during the day. The lines were so close that the utmost care was required to obtain a satisfactory shot without an exposure which was almost certainly fatal. Nevertheless, the gallant men of the Vermont company managed to use up the one hundred rounds of cartridges with which they were supplied long before the day was over. Capt. Merriman, foreseeing this, had directed Sergt. Cassius Peck to procure a fresh supply.

It was a service of grave danger, but taking two haversacks the sergeant succeeded in safely passing twice over the dangerous ground and thus enabled the company to hold its threatened lines. Many men in the company fired as many as two hundred rounds on this day, and at its close the rifles were so choked with dirt and dust, and so heated with the rapid and continuous firing, as to be almost unserviceable.

The company suffered a severe loss at this place by the death of Corporal Charles B. Mead, who was shot through the head and instantly killed. Corporal Mead was one of the recruits who joined in the autumn of 1862, and had been constantly with the company and constantly on duty ever since, except while recovering from a former wound received at Gettysburg. He was one of two brothers who enlisted at the same time, the other, Carlos E. Mead, having been himself wounded. He was a young man of rare promise, and his early death brought sadness, not only to his comrades in the field, but to a large circle of friends at home. He had kept a daily record of events in the form of a diary during his entire period of service, to which the writer of these lines has had access, and from which he has obtained valuable information and assistance in his work.

Henry E. Barnum was also mortally wounded, and died on the fourteenth of the following month, while John Quinlan received a severe wound. Quinlan, however, recovered and served his enlistment to the close of the war. Sergt.-Major Jacobs, formerly of Co. G, who served with Co. F on this day, was also mortally wounded.

The company was relieved at night and retired to the rear for a well earned rest, to be engaged the next day in the sharp engagement around the Hare house. Their position here, however, was less exposed and their service less arduous. The Hare house had but lately been vacated by its former occupants, a wealthy and influential Virginia family, who had left so suddenly as to have abandoned nearly everything that the house contained. The windows of the basement opened full on the rebel works and rifle pits, the latter within point bank range, and here the sharpshooters, seated at ease in the fine mahogany chairs of the late owner, took careful aim at his friends in his own garden. They boiled their coffee, and cooked their rashers of pork, on his cooking range, over fires started and fed with articles taken from his elegant apartments, not, it is to be feared, originally intended for fuel, and ate them on his dining table. There was, however, no vandalism, no wanton destruction of property for the mere sake of destruction in all this. The house and its contents were doomed in any event, and the slight havoc worked by the sharpshooters only anticipated by a few hours what must come in a more complete form later. The shooting here was at very short range, and correspondingly accurate. As an Alabama rifleman, who was taken prisoner, remarked, "It was only necessary to hold up your hand to get a furlough, and you were lucky if you could get to the rear without an extension."

Silas Giddings was wounded here. Giddings had been a friend

and schoolmate of the Meads, and had enlisted at the same time. Thus of the three friends two were severely wounded and one was dead. During the day Birney's division had made an assault on the main rebel line to the left of the Hare house which had been repulsed with severe loss. The wounded were left on the field, some of them close under the enemy's works. They lay in plain sight during the hours of daylight, but it was impossible to help them.

When darkness came on, however, Capt. Merriman, slinging half a dozen canteens over his shoulder, crept out onto the field and spent half the night in caring for the poor fellows whose sufferings during the day had so touched his sympathies.

The 19th, 20th and 21st of June were spent at this place, sharpshooting constantly going on. On the twentieth Corporal Edward Lyman received a wound of which he died on the twenty-fifth. Corporal Lyman was one of the original members of the company; was promoted corporal on the 15th of August, 1863, and had long been a member of the colour guard of the regiment, having been selected for that position for his distinguished courage and coolness on many fields. Some times during these days a temporary truce would be agreed upon between the opposing pickets, generally for the purpose of boiling coffee or preparing food. Half an hour perhaps would be the limit of time agreed upon; but whatever it was, the truce was scrupulously observed. When some one called "time," however, it behoved every man to take cover instantly.

Upon one occasion a rebel rifleman was slow to respond to the warning—in fact he appeared to think himself out of sight; while all others hurried to their posts he alone sat quietly blowing his hot coffee and munching his hard-tack. It so happened, however, that he was in plain sight of a sharpshooter less blood-thirsty than some others, who thought it only fair to give him one more warning, therefore he called out, "I say, Johnny, time is up, get into your hole."

"All right," responded the cool rebel still blowing away at his hot cup.

"Just hold that cup still," said the sharpshooter, "and I will show you whether it is all right or not."

By this time the fellow began to suspect that he was indeed visible, and holding his cup still for an instant while he looked up, he afforded the Union marksman the opportunity he was waiting for. A rapid sight and the sharp's bullet knocked the coffee cup far out of its owner's reach and left it in such a condition that it could never serve a useful purpose again. The surprised rebel made haste to get under cover, pursued by the laughter and jeers of his own comrades as well as those of the sharpshooters. Thus men played practical jokes on each other at one moment, and the next were seeking to do each other mortal harm.

The various assaults having failed to force the enemy from any considerable portion of the defences of Petersburg, it was determined by the federal commanders to extend again to the left, with the intent to cut off, one by one, the avenues by which supplies might be brought to the enemy from the South; and on the twenty-first the Second Corps, now under Gen. Birney (Gen. Hancock being disabled by the reopening of an old wound), in company with the Fifth and Sixth Corps, moved to the left and took up a position with its right on the Jerusalem plank road. The Sixth Corps, which was to have prolonged the line to the left, not arriving in position as early as was expected, the enemy took instant advantage of the opportunity and, penetrating to the rear of the exposed left of the Second Corps, commenced a furious attack. Thus surprised, the entire left division gave way in disorder and retreated towards the right, thus uncovering the left of Mott's division, which was next in line, which in its turn was thrown into confusion. The sharpshooters, who had been skirmishing in advance of the left, had, of course, no option; they were compelled to retire with their supports or submit to capture. They fell back slowly and in good order, however, gradually working themselves into a position to partially check the advancing rebels and afford a scanty space of time in which the disordered mass might rally and reform. In this movement they were gallantly supported by the Fifth Michigan Volunteers by whose assistance they were, at last, enabled to bring the rebels to a halt; not, however, until they had captured some seventeen hundred men and four guns from the corps. The company again suffered heavy loss in this affair.

Barney Leddy and Peter Lafflin were killed on the field; Watson P. Morgan was wounded and taken prisoner; Sergt. Grover was badly wounded by a rifle ball through the thigh, and David Clark received a severe wound. Morgan was a young but able and gallant soldier; he had previously been wounded at Kelly's ford, but returned to his company to be again wounded, and to experience the additional misfortune of being made a prisoner. He was exchanged soon after, but subsequently died from the effect of his wound. Sergt. Grover had also previously been wounded at Gettysburg, where he had been promoted for gallantry and good conduct. Clark recovered to reenlist upon the expiration of his term of service, and served to the close of the war.

Of the forty-seven men who had been with the company since it crossed the Rapidan only ten were left for duty—thirty-five had been killed or wounded, and two had been captured unwounded.

From this time to the 26th of July the company were employed, with short intervals for rest, on the picket line, here and there as occasion demanded their services, but without important incident. Active operations having now continued so long in this particular quarter as to afford room for hope that the rebels might be caught napping on the north bank of the James, Gen. Grant determined to send a large force in that direction to co-operate with the Army of the James, hoping to take the enemy by surprise and, by a sudden dash, perhaps to capture the capital of the Confederacy before its real defenders could get information of the danger. With this view he detached the Second Corps and two divisions of cavalry to attempt it.

The troops marched at one o'clock on the afternoon of the twenty-sixth, and at two o'clock on the morning of the twenty-seventh the corps crossed the James by a pontoon bridge at Jones'

Landing. Passing rapidly to the north, in rear of the lines held by the Tenth Corps (belonging to the Army of the James), the troops faced to the west and were soon confronting the enemy in position. The sharpshooters were deployed and advanced in skirmishing order across an open and level tract of land known locally as Strawberry Plains.

The advancing line was heavily supported and drove the enemy steadily until they were forced back into their works, when, with a grand dash, sharpshooters, supports and all in one rushing mass, swept up to and over the rebel works, capturing in the charge four guns and some seven hundred prisoners. Notwithstanding this success, the enemy were found to have been so heavily reinforced by troops from the Petersburg lines—who could be transferred by railroad, while the Union forces were compelled to march—that the full object of the movement could not be attained. The captured works were held, however, while the cavalry, moving still further north, destroyed the railroads and bridges north of the city, and returned to the vicinity of Deep Bottom, where the corps returned by a night march to their former position in front of Petersburg, resting for a few hours by the way on the field of their battle of the 18th of July.

The regiment lay in camp until the 12th of August, engaged in the usual routine of picket duly and sharpshooting, but without unusually hard service. Indeed, what would once have been called by them active employment was now enjoyed as a season of grateful repose, so constantly had they been engaged in bloody battle since crossing the Rapidan.

On the 12th of August the bugle sounded the general once more, and with knapsacks packed, blankets strapped, haversacks and cartridge boxes filled, the one hundred and sixty men who now represented what had once been the First Regiment of United States Sharpshooters, marched with their division towards City Point.

Rumours were rife as to their destination—some said Washington; some said a southern seaport, while some maintained that the objective point was Chicago, where they were wanted to maintain order during the coming democratic convention. At City Point they were embarked on steam transports and headed down the river. The wisest guessers were now really puzzled, and the prophet who foretold Chicago had as many chances in his favour as any of his fellows. A few miles down the river, and the fleet of laden steamers came to an anchor, and lay quiet for some hours. The rest, cleanliness, and cool, refreshing breezes from the river, were very grateful to the tired soldiers so long accustomed to the dirt and dust of the rifle pits.

Soon after dark the anchors were got up and the heads of the steamers turned again up stream. Now all was plain, another secret movement was planned, and at daylight on the morning of the fourteenth the troops landed at the scene of their crossing on the 26th of July at Deep Bottom.

Moving out toward the enemy severe skirmishing took place, but no engagement of a general character occurred on that day. On the fifteenth they were detached from the Second, and ordered to the Tenth Corps, now commanded by their former division commander, Gen. Birney, and at his especial request. Moving out at the head of the column they found themselves in the early afternoon the extreme right of the army, and in front of the enemy at a little stream known as Deep Run, or Four Mile Creek. Deploying under the personal direction of Gen. Birney they advanced toward a wooded ridge on which they found the rebel skirmishers in force, and evidently determined to stay. In the language of Capt. Merriman, who must be accepted as authority, "It was the hardest skirmish line to start that Co. F ever struck." But Co. F was rarely refused when it demanded a right of way and was opposed by nothing but a skirmish line; and on this occasion, as on many former ones, their steady pressure and cool firing prevailed at last, and after more than an hour the rebels yielded the ground.

On the sixteenth more severe fighting took place with serious loss to the regiment, but Co. F escaped without loss—in fact there was hardly enough left of the company to lose. Col. Craig, commanding the brigade to which they were attached, was killed, and Capt. Andrews of Co. E, Capt. Aschmann of Co. A. and Lieut. Tyler of Co. I were wounded. Thus this movement ended, as had the former one, with no decisive result so far as the participants could see. A few guns had been captured, a few rebels killed, and a corresponding loss had been suffered by the federals; but who could tell what important effect on the great field of action, extending from the Mississippi to the Atlantic, this apparently abortive movement was intended to have?

The men were beginning to understand that marches and battles were not always for immediate effect at the point of contact; and so they marched and fought as they were ordered; winning if they could, and accepting defeat if they must, but with a growing confidence that the end was near.

On the seventeenth they rejoined their proper corps and marched again toward the James, leaving Lieut. Kinsman in charge of a party who, under a flag of truce, was caring for the wounded.

The corps re-crossed the James on the night of the nineteenth and resumed a place in the lines of Petersburg, relieving the Fifth Corps who moved to the left to try to seize and hold the Weldon railroad, the attempt on which had been abandoned since the battle on the Jerusalem plank road on the 22nd of July. On the twentieth, companies C and A, whose term of service had expired, were discharged. In Co. C only five, and in Co. A. only eleven of the original members were left to be mustered out. The terrible exposures of three years of fighting had done their perfect work on them, and the little band who answered to the roll call on that day had little resemblance to the sturdy line that had raised their hands as they took the oath only three years before. The regiment was on the eve of dissolution, since other companies were soon to reach the end of their enlistment and might soon be expected to leave the service. Indeed, the company whose history we have followed so long, would be entitled to its discharge on the 12th of September, now only twenty-three days off.

The departure of Co. A was made more sad from the fact

that they took with them their wounded captain, who had lost a leg in the battle at Deep Run on the fifteenth. Capt. Aschmann had been with the company from its organization, and had participated with distinguished gallantry in all the battles in which it had been engaged, escaping without a wound, only to lose his leg in the last fight, and only five days before he would be entitled to his honourable discharge. It seemed a hard fate. In Co. F great excitement existed in consequence of the near approach of the time when they, also, might honourably doff the green uniforms which had so long been worn as a distinctive mark of their organization, and turn their faces homeward, once more to become sober citizens in the peaceful and prosperous North—that North which they had fought so long and so hard to preserve in its peace and prosperity. Many and frequent were the discussions around the camp fire as to whether it was better to leave the service or to reenlist.

It was now plain that the days of the rebellion were numbered, and that the end was at hand. It was evident to these veterans, however, that a few more desperate battles must be fought before the end was finally reached. They ardently desired to be present at the final surrender and share the triumph they had suffered so much to assure. On the other hand they as ardently longed to resume their places in those home circles which they had left to take up arms, only that the country and the flag, which they so honoured and loved, might be preserved to their children, and their children's children, forever. They felt that they had done all that duty required of them, that they had honourably served their term, and that they might safely leave it to those who had entered the service later to finish the work which they had so far completed. They felt, also, that they should leave behind them an honourable record, on which no stain rested, and second to that of no body of men in the army.

There were left of the original one hundred and three men who had been mustered into the United States service only twenty-five present and absent. Of these, six, namely, David Clark, Jas. H. Guthrie, Sam'l J. Williams, Stephen B. Flanders,

John Kanaan and Lewis J. Allen, had reenlisted. The remainder, nineteen in number, as follows, elected to take their honourable discharge:

C. D. Merriman	Spafford A. Wright
W. C. Kent	Eugene Payne
Fitz Green Halleck	H. E. Kinsman
Wm. McKeever	Almon D. Griffin
Watson N. Sprague	Jas. M. Thompson
W. W. Cutting	David O. Daggett
H. B. Wilder	Curtiss P. Kimberly
Cassius Peck	Edwin E. Robinson
E. F. Stevens	Thos. H. Turnbull
Geo. H. Ellis	

Of these, nine only were present with the company to be mustered out. The remaining six were absent, sick or wounded, or on detached service.

The few remaining days were destined, however, to be full of excitement and danger. It seemed to the men that their division commander, aware of the fact that he was soon to lose them, was determined to use them to the best advantage while he had them. They were kept constantly engaged during the hours of daylight, skirmishing and engaged in the rifle pits. On the 21st of August they drove the rebels from a rifle pit in their front, capturing forty prisoners, just four times as many as there were men in their own ranks. From this date until September 10th they were engaged every day on the picket line. On that day, with other companies, they were ordered to occupy what had been, by consent, neutral ground surrounding a well from which both parties had drawn water, and where rebel and Unionist often met unarmed and exchanged gossip. It seemed a pity to spoil so friendly an arrangement, but orders must be obeyed, and soon after daylight the sharpshooters dashed out of their lines and occupied the ground which they proceeded to fortify, capturing eighty-five surprised, but not on the whole displeased, rebels.

The enemy did not relish being deprived of the opportunity

of getting water from this place, and on that day and the next made repeated effort to repossess it, but without avail. Carlos E. Mead received his second wound in repulsing one of these attempts. At last the day arrived when they might claim to have fulfilled on their part the engagement which they had entered into with Uncle Sam three years before, and on the thirteenth the men present took their final discharge and bade farewell to all the "Pomp and circumstance of glorious war." They were destined, however, to have one more opportunity to show their skill even on this last day of their service, for even while they were preparing for their leave taking a sharp exchange of shots took place, in which the departing Vermonters paid their last compliments to the enemy whom they had so often fought, and during which A. W. Bemis, a recruit of 1862, was wounded. At last all was over; reluctantly turning in their trusty rifles, to which they had become attached by long companionship in many scenes of danger and death, they answered to the last roll call and, bidding an affectionate adieu to their comrades who were to remain, they turned their faces toward City Point and home.

The small remnant of the company kept up an organization under Sergt. Cunningham, and was heavily engaged on the 27th of October in the battle of Burgess Mill, which resulted from Grant's attempt on the South Side railroad. The few men left fought with their accustomed gallantry, losing Daniel E. Bessie and Charles Danforth, killed in action, and Volney W. Jencks and Jay S. Percy, wounded and left on the field.

The little squad, now reduced to almost nothing, were again engaged on the 1st of November, when they suffered the loss of still another comrade, Friend Weeks, who was mortally wounded and died on the seventeenth of the same month. On the 23rd of December the few men left of the once strong and gallant company were transferred to Co. E of the Second Sharpshooters, and Co. F ceased to exist as an organization. With Co. E the men so transferred participated in the affair at Hatcher's Run on the 15th of December, and at other points along the line. On the 25th of February, 1865, the consolidated battalion of sharpshooters being reduced to a mere skeleton, was broken up and its members transferred to other regiments, the Vermonters being assigned to Co. G, Fourth Vermont Volunteers, with which company they served until the close of the war.

On the 16th of February, the division commander, Gen. De Trobriand, under whom they had served for nearly two years, and who knew them better, probably, than any general officer of the army, issued the following complimentary order:

Headquarters
3rd Div. 2nd Army Corps
February 16, 1865
General Order No. 12
The United States Sharpshooters, including the first and second consolidated battalions, being about to be broken up as a distinct organization in compliance with orders from the War Department, the brigadier-general commanding the division will not take leave of them without acknowledging their good and efficient service during about three years in the field. The United States Sharpshooters leave behind them a glorious record in the Army of the Potomac since the first operations against Yorktown in 1862 up to Hatcher's Run, and few are the battles or engagements where they did not make their mark. The brigadier-general commanding, who had them under his command during most of the campaigns of 1863 and 1864, would be the last to forget their brave deeds during that period, and he feels assured that in the different organizations to which they may belong severally, officers and men will show themselves worthy of their old reputation; with them the past will answer for the future.
By command of Brig.-Gen. R. De Trobriand.
W. K. Driver
A. A. G.

It was a handsome compliment on the part of the commander, well deserved and heartily bestowed. The history of Co. F would not be complete, or do justice to the memories of the brave men who died in its ranks, or to the gallant few yet living, without a record of the names of those who so freely shed their blood, in the conflict for the Union.

In all thirty-two of its members died of wounds received in action, of whom twenty-one were killed on the field while eleven died in the hospital from the effects of their wounds. Their names are as follows:

A. H. Cooper	Jay S. Percy
W. J. Domag	Jacob Lacoy
Thos. H. Brown	Caspar B. Kent
Dan'l E. Bessie	W. F. Dawson
M. W. Wilson	Alvin Babcock
Watson P. Morgan	Volney W. Jencks
David W. French	Edw'd Trask
Henry Mattocks	Jos. Bickford
Peter Lafflin	Chas. Danforth
A. C. Cross	Jno. Bowen
Friend Weeks	William Wells.
E. M. Hosmer	Joseph Hagan
Barney Leddy	Jas. A. Read
Edw'd Lyman	Pat'k Murray
E. A. Giddings	Chas. B. Mead
B. W. Jordan	Henry E. Barnum

The wounded who recovered and again reported for duty number forty-five. The names are given here as second in honourable recollection only to those who died on the field. The list will be found to contain the names of several who were subsequently killed, or died of wounds received on other fields:

C. M. Jordan	Wm. McKeever
Dustin K. Bareau	Edward Lyman
John Quinlan	L. D. Grover (twice)
Sam'l Williams	C. W. Peck
C. W. Seaton	W. C. Kent
W. H. Blake	Barney Leddy
Jno. Monahan	Chas. B. Mead
A. J. Cross	Jno. C. Page

H. E. Kinsman	Henry Mattocks
Almon D. Griffin (twice)	Silas Giddings
Carlos E. Mead (twice)	Geo. Woolly
E. H. Himes	Jacob S. Bailey (twice)
Ai Brown	S. M. Butler
Martin C. Laffie	W. H. Leach
Fitz Greene Halleck	Eugene Payne
Spafford A. Wright	J. E. Chase
A. W. Bemis	Benjamin Billings
Brigham Buswell	E. M. Hosmer
Watson P. Morgan	M. Cunningham (twice)
Amos A. Smith	David Clark
Lewis J. Allen	H. J. Peck
Edward Trask	Edw'd Jackson
Sherod Brown.	

Thus out of a total of one hundred and seventy-seven men, including all recruits actually mustered into the United States service (for it must be remembered that thirteen of the one hundred and sixteen men who were mustered by the state mustering officer at Randolph, and charged against the company on the rolls, were discharged at Washington to reduce the number to the legal requirement of one hundred and three officers and men, thirty-two, or more than eighteen per cent, died of wounds; while the killed and wounded taken together number seventy-seven, or forty-three and one-half per cent of the whole.

The record shows the severe and dangerous nature of the service performed by these men, and on it they may safely rest, certain that a grateful country will honour their memories, even as it does those of their comrades who fought in the ranks of other and larger organizations.

Conclusion

You can have ten descriptions of a battle, or plans of a campaign, sooner than one glimpse at the unthought of details of a soldier's life.

The history of Co. F is finished, and is far from satisfactory to the writer. Originally undertaken for the purpose of supplying the Hon. G. G. Benedict, state military historian, with material for such a brief record as he could afford room for in his history of the Vermont troops in the war of the rebellion, it has grown far beyond what was intended at the outset, and far beyond what would be proper for him to publish in such a work as he is charged with. It should have been undertaken by some other person than myself; by someone more intimately and longer acquainted with the company in the field: by someone whose personal recollection of the detail of its daily doings is more exact than mine can possibly be; for the history of so small a portion of a great army as a company is, should be a personal history of the men who composed it. The record of a company is mainly made up of the every day scenes and every day gossip about its company kitchen and its company street. With these matters the writer does not profess to be, or to have been, familiar.

The work has, therefore, become more of a description of campaigns and of battles, and more a history of the regiment to which it was attached, I fear, than of the company. Such as it is, however, its preparation has been a labour of love, and it is published with the earnest hope that it may serve at least to keep warm in the hearts of the survivors the memories of those who marched with them in 1861, and whose graves mark every battle field whereon the Army of the Potomac fought.

Wm. Y. W. R.

LEONAUR

ALSO FROM LEONAUR
AVAILABLE IN SOFTCOVER OR HARDCOVER WITH DUST JACKET

AFGHANISTAN: THE BELEAGUERED BRIGADE *by G. R. Gleig*—An Account of Sale's Brigade During the First Afghan War.

IN THE RANKS OF THE C. I. V *by Erskine Childers*—With the City Imperial Volunteer Battery (Honourable Artillery Company) in the Second Boer War.

THE BENGAL NATIVE ARMY *by F. G. Cardew*—An Invaluable Reference Resource.

THE 7TH (QUEEN'S OWN) HUSSARS: Volume 4—1688-1914 *by C. R. B. Barrett*—Uniforms, Equipment, Weapons, Traditions, the Services of Notable Officers and Men & the Appendices to All Volumes—Volume 4: 1688-1914.

THE SWORD OF THE CROWN *by Eric W. Sheppard*—A History of the British Army to 1914.

THE 7TH (QUEEN'S OWN) HUSSARS: Volume 3—1818-1914 *by C. R. B. Barrett*—On Campaign During the Canadian Rebellion, the Indian Mutiny, the Sudan, Matabeleland, Mashonaland and the Boer War Volume 3: 1818-1914.

THE KHARTOUM CAMPAIGN *by Bennet Burleigh*—A Special Correspondent's View of the Reconquest of the Sudan by British and Egyptian Forces under Kitchener—1898.

EL PUCHERO *by Richard McSherry*—The Letters of a Surgeon of Volunteers During Scott's Campaign of the American-Mexican War 1847-1848.

RIFLEMAN SAHIB *by E. Maude*—The Recollections of an Officer of the Bombay Rifles During the Southern Mahratta Campaign, Second Sikh War, Persian Campaign and Indian Mutiny.

THE KING'S HUSSAR *by Edwin Mole*—The Recollections of a 14th (King's) Hussar During the Victorian Era.

JOHN COMPANY'S CAVALRYMAN *by William Johnson*—The Experiences of a British Soldier in the Crimea, the Persian Campaign and the Indian Mutiny.

COLENSO & DURNFORD'S ZULU WAR *by Frances E. Colenso & Edward Durnford*—The first and possibly the most important history of the Zulu War.

U. S. DRAGOON *by Samuel E. Chamberlain*—Experiences in the Mexican War 1846-48 and on the South Western Frontier.

LEONAUR

ALSO FROM LEONAUR
AVAILABLE IN SOFTCOVER OR HARDCOVER WITH DUST JACKET

THE FALL OF THE MOGHUL EMPIRE OF HINDUSTAN *by H. G. Keene*—By the beginning of the nineteenth century, as British and Indian armies under Lake and Wellesley dominated the scene, a little over half a century of conflict brought the Moghul Empire to its knees.

LADY SALE'S AFGHANISTAN *by Florentia Sale*—An Indomitable Victorian Lady's Account of the Retreat from Kabul During the First Afghan War.

THE CAMPAIGN OF MAGENTA AND SOLFERINO 1859 *by Harold Carmichael Wylly*—The Decisive Conflict for the Unification of Italy.

FRENCH'S CAVALRY CAMPAIGN *by J. G. Maydon*—A Special Correspondent's View of British Army Mounted Troops During the Boer War.

CAVALRY AT WATERLOO *by Sir Evelyn Wood*—British Mounted Troops During the Campaign of 1815.

THE SUBALTERN *by George Robert Gleig*—The Experiences of an Officer of the 85th Light Infantry During the Peninsular War.

NAPOLEON AT BAY, 1814 *by F. Loraine Petre*—The Campaigns to the Fall of the First Empire.

NAPOLEON AND THE CAMPAIGN OF 1806 *by Colonel Vachée*—The Napoleonic Method of Organisation and Command to the Battles of Jena & Auerstädt.

THE COMPLETE ADVENTURES IN THE CONNAUGHT RANGERS *by William Grattan*—The 88th Regiment during the Napoleonic Wars by a Serving Officer.

BUGLER AND OFFICER OF THE RIFLES *by William Green & Harry Smith*—With the 95th (Rifles) during the Peninsular & Waterloo Campaigns of the Napoleonic Wars.

NAPOLEONIC WAR STORIES *by Sir Arthur Quiller-Couch*—Tales of soldiers, spies, battles & sieges from the Peninsular & Waterloo campaigns.

CAPTAIN OF THE 95TH (RIFLES) *by Jonathan Leach*—An officer of Wellington's sharpshooters during the Peninsular, South of France and Waterloo campaigns of the Napoleonic wars.

RIFLEMAN COSTELLO *by Edward Costello*—The adventures of a soldier of the 95th (Rifles) in the Peninsular & Waterloo Campaigns of the Napoleonic wars.

LEONAUR

ALSO FROM LEONAUR
AVAILABLE IN SOFTCOVER OR HARDCOVER WITH DUST JACKET

OFFICERS & GENTLEMEN *by Peter Hawker & William Graham*—Two Accounts of British Officers During the Peninsula War: Officer of Light Dragoons by Peter Hawker & Campaign in Portugal and Spain by William Graham .

THE WALCHEREN EXPEDITION *by Anonymous*—The Experiences of a British Officer of the 81st Regt. During the Campaign in the Low Countries of 1809.

LADIES OF WATERLOO *by Charlotte A. Eaton, Magdalene de Lancey & Juana Smith*—The Experiences of Three Women During the Campaign of 1815: Waterloo Days by Charlotte A. Eaton, A Week at Waterloo by Magdalene de Lancey & Juana's Story by Juana Smith.

JOURNAL OF AN OFFICER IN THE KING'S GERMAN LEGION *by John Frederick Hering*—Recollections of Campaigning During the Napoleonic Wars.

JOURNAL OF AN ARMY SURGEON IN THE PENINSULAR WAR *by Charles Boutflower*—The Recollections of a British Army Medical Man on Campaign During the Napoleonic Wars.

ON CAMPAIGN WITH MOORE AND WELLINGTON *by Anthony Hamilton*—The Experiences of a Soldier of the 43rd Regiment During the Peninsular War.

THE ROAD TO AUSTERLITZ *by R. G. Burton*—Napoleon's Campaign of 1805.

SOLDIERS OF NAPOLEON *by A. J. Doisy De Villargennes & Arthur Chuquet*—The Experiences of the Men of the French First Empire: Under the Eagles by A. J. Doisy De Villargennes & Voices of 1812 by Arthur Chuquet .

INVASION OF FRANCE, 1814 *by F. W. O. Maycock*—The Final Battles of the Napoleonic First Empire.

LEIPZIG—A CONFLICT OF TITANS *by Frederic Shoberl*—A Personal Experience of the 'Battle of the Nations' During the Napoleonic Wars, October 14th-19th, 1813.

SLASHERS *by Charles Cadell*—The Campaigns of the 28th Regiment of Foot During the Napoleonic Wars by a Serving Officer.

BATTLE IMPERIAL *by Charles William Vane*—The Campaigns in Germany & France for the Defeat of Napoleon 1813-1814.

SWIFT & BOLD *by Gibbes Rigaud*—The 60th Rifles During the Peninsula War.

LEONAUR

ALSO FROM LEONAUR
AVAILABLE IN SOFTCOVER OR HARDCOVER WITH DUST JACKET

ADVENTURES OF A YOUNG RIFLEMAN by Johann Christian Maempel—The Experiences of a Saxon in the French & British Armies During the Napoleonic Wars.

THE HUSSAR by Norbert Landsheit & G. R. Gleig—A German Cavalryman in British Service Throughout the Napoleonic Wars.

RECOLLECTIONS OF THE PENINSULA by Moyle Sherer—An Officer of the 34th Regiment of Foot—'The Cumberland Gentlemen'—on Campaign Against Napoleon's French Army in Spain.

MARINE OF REVOLUTION & CONSULATE by Moreau de Jonnès—The Recollections of a French Soldier of the Revolutionary Wars 1791-1804.

GENTLEMEN IN RED by John Dobbs & Robert Knowles—Two Accounts of British Infantry Officers During the Peninsular War Recollections of an Old 52nd Man by John Dobbs An Officer of Fusiliers by Robert Knowles.

CORPORAL BROWN'S CAMPAIGNS IN THE LOW COUNTRIES by Robert Brown—Recollections of a Coldstream Guard in the Early Campaigns Against Revolutionary France 1793-1795.

THE 7TH (QUEENS OWN) HUSSARS: Volume 2—1793-1815 by C. R. B. Barrett—During the Campaigns in the Low Countries & the Peninsula and Waterloo Campaigns of the Napoleonic Wars. Volume 2: 1793-1815.

THE MARENGO CAMPAIGN 1800 by Herbert H. Sargent—The Victory that Completed the Austrian Defeat in Italy.

DONALDSON OF THE 94TH—SCOTS BRIGADE by Joseph Donaldson—The Recollections of a Soldier During the Peninsula & South of France Campaigns of the Napoleonic Wars.

A CONSCRIPT FOR EMPIRE by Philippe as told to Johann Christian Maempel—The Experiences of a Young German Conscript During the Napoleonic Wars.

JOURNAL OF THE CAMPAIGN OF 1815 by Alexander Cavalié Mercer—The Experiences of an Officer of the Royal Horse Artillery During the Waterloo Campaign.

NAPOLEON'S CAMPAIGNS IN POLAND 1806-7 by Robert Wilson—The campaign in Poland from the Russian side of the conflict.

ALSO FROM LEONAUR
AVAILABLE IN SOFTCOVER OR HARDCOVER WITH DUST JACKET

COLBORNE: A SINGULAR TALENT FOR WAR *by John Colborne*—The Napoleonic Wars Career of One of Wellington's Most Highly Valued Officers in Egypt, Holland, Italy, the Peninsula and at Waterloo.

NAPOLEON'S RUSSIAN CAMPAIGN *by Philippe Henri de Segur*—The Invasion, Battles and Retreat by an Aide-de-Camp on the Emperor's Staff.

WITH THE LIGHT DIVISION *by John H. Cooke*—The Experiences of an Officer of the 43rd Light Infantry in the Peninsula and South of France During the Napoleonic Wars.

WELLINGTON AND THE PYRENEES CAMPAIGN VOLUME I: FROM VITORIA TO THE BIDASSOA *by F. C. Beatson*—The final phase of the campaign in the Iberian Peninsula.

WELLINGTON AND THE INVASION OF FRANCE VOLUME II: THE BIDASSOA TO THE BATTLE OF THE NIVELLE *by F. C. Beatson*—The final phase of the campaign in the Iberian Peninsula.

WELLINGTON AND THE FALL OF FRANCE VOLUME III: THE GAVES AND THE BATTLE OF ORTHEZ *by F. C. Beatson*—The final phase of the campaign in the Iberian Peninsula.

NAPOLEON'S IMPERIAL GUARD: FROM MARENGO TO WATERLOO *by J. T. Headley*—The story of Napoleon's Imperial Guard and the men who commanded them.

BATTLES & SIEGES OF THE PENINSULAR WAR *by W. H. Fitchett*—Corunna, Busaco, Albuera, Ciudad Rodrigo, Badajos, Salamanca, San Sebastian & Others.

SERGEANT GUILLEMARD: THE MAN WHO SHOT NELSON? *by Robert Guillemard*—A Soldier of the Infantry of the French Army of Napoleon on Campaign Throughout Europe.

WITH THE GUARDS ACROSS THE PYRENEES *by Robert Batty*—The Experiences of a British Officer of Wellington's Army During the Battles for the Fall of Napoleonic France, 1813 .

A STAFF OFFICER IN THE PENINSULA *by E. W. Buckham*—An Officer of the British Staff Corps Cavalry During the Peninsula Campaign of the Napoleonic Wars.

THE LEIPZIG CAMPAIGN: 1813—NAPOLEON AND THE "BATTLE OF THE NATIONS" *by F. N. Maude*—Colonel Maude's analysis of Napoleon's campaign of 1813 around Leipzig.

LEONAUR

ALSO FROM LEONAUR
AVAILABLE IN SOFTCOVER OR HARDCOVER WITH DUST JACKET

BUGEAUD: A PACK WITH A BATON by *Thomas Robert Bugeaud*—The Early Campaigns of a Soldier of Napoleon's Army Who Would Become a Marshal of France.

WATERLOO RECOLLECTIONS by *Frederick Llewellyn*—Rare First Hand Accounts, Letters, Reports and Retellings from the Campaign of 1815.

SERGEANT NICOL by *Daniel Nicol*—The Experiences of a Gordon Highlander During the Napoleonic Wars in Egypt, the Peninsula and France.

THE JENA CAMPAIGN: 1806 by *F. N. Maude*—The Twin Battles of Jena & Auerstadt Between Napoleon's French and the Prussian Army.

PRIVATE O'NEIL by *Charles O'Neil*—The recollections of an Irish Rogue of H. M. 28th Regt.—The Slashers—during the Peninsula & Waterloo campaigns of the Napoleonic war.

ROYAL HIGHLANDER by *James Anton*—A soldier of H.M 42nd (Royal) Highlanders during the Peninsular, South of France & Waterloo Campaigns of the Napoleonic Wars.

CAPTAIN BLAZE by *Elzéar Blaze*—Life in Napoleons Army.

LEJEUNE VOLUME 1 by *Louis-François Lejeune*—The Napoleonic Wars through the Experiences of an Officer on Berthier's Staff.

LEJEUNE VOLUME 2 by *Louis-François Lejeune*—The Napoleonic Wars through the Experiences of an Officer on Berthier's Staff.

CAPTAIN COIGNET by *Jean-Roch Coignet*—A Soldier of Napoleon's Imperial Guard from the Italian Campaign to Russia and Waterloo.

FUSILIER COOPER by *John S. Cooper*—Experiences in the 7th (Royal) Fusiliers During the Peninsular Campaign of the Napoleonic Wars and the American Campaign to New Orleans.

FIGHTING NAPOLEON'S EMPIRE by *Joseph Anderson*—The Campaigns of a British Infantryman in Italy, Egypt, the Peninsular & the West Indies During the Napoleonic Wars.

CHASSEUR BARRES by *Jean-Baptiste Barres*—The experiences of a French Infantryman of the Imperial Guard at Austerlitz, Jena, Eylau, Friedland, in the Peninsular, Lutzen, Bautzen, Zinnwald and Hanau during the Napoleonic Wars.

LEONAUR

ALSO FROM LEONAUR
AVAILABLE IN SOFTCOVER OR HARDCOVER WITH DUST JACKET

LIFE IN THE ARMY OF NORTHERN VIRGINIA *by Carlton McCarthy*—The Observations of a Confederate Artilleryman of Cutshaw's Battalion During the American Civil War 1861-1865.

HISTORY OF THE CAVALRY OF THE ARMY OF THE POTOMAC *by Charles D. Rhodes*—Including Pope's Army of Virginia and the Cavalry Operations in West Virginia During the American Civil War.

CAMP-FIRE AND COTTON-FIELD *by Thomas W. Knox*—A New York Herald Correspondent's View of the American Civil War.

SERGEANT STILLWELL *by Leander Stillwell* —The Experiences of a Union Army Soldier of the 61st Illinois Infantry During the American Civil War.

STONEWALL'S CANNONEER *by Edward A. Moore*—Experiences with the Rockbridge Artillery, Confederate Army of Northern Virginia, During the American Civil War.

THE SIXTH CORPS *by George Stevens*—The Army of the Potomac, Union Army, During the American Civil War.

THE RAILROAD RAIDERS *by William Pittenger*—An Ohio Volunteers Recollections of the Andrews Raid to Disrupt the Confederate Railroad in Georgia During the American Civil War.

CITIZEN SOLDIER *by John Beatty*—An Account of the American Civil War by a Union Infantry Officer of Ohio Volunteers Who Became a Brigadier General.

COX: PERSONAL RECOLLECTIONS OF THE CIVIL WAR--VOLUME 1 *by Jacob Dolson Cox*—West Virginia, Kanawha Valley, Gauley Bridge, Cotton Mountain, South Mountain, Antietam, the Morgan Raid & the East Tennessee Campaign.

COX: PERSONAL RECOLLECTIONS OF THE CIVIL WAR--VOLUME 2 *by Jacob Dolson Cox*—Siege of Knoxville, East Tennessee, Atlanta Campaign, the Nashville Campaign & the North Carolina Campaign.

KERSHAW'S BRIGADE VOLUME 1 *by D. Augustus Dickert*—Manassas, Seven Pines, Sharpsburg (Antietam), Fredericksburg, Chancellorsville, Gettysburg, Chickamauga, Chattanooga, Fort Sanders & Bean Station.

KERSHAW'S BRIGADE VOLUME 2 *by D. Augustus Dickert*—At the wilderness, Cold Harbour, Petersburg, The Shenandoah Valley and Cedar Creek..

www.ingramcontent.com/pod-product-compliance
Lightning Source LLC
Chambersburg PA
CBHW021107090426

42738CB00006B/541